Springer Theses

Recognizing Outstanding Ph.D. Research

For further volumes:
http://www.springer.com/series/8790

Aims and Scope

The series "Springer Theses" brings together a selection of the very best Ph.D. theses from around the world and across the physical sciences. Nominated and endorsed by two recognized specialists, each published volume has been selected for its scientific excellence and the high impact of its contents for the pertinent field of research. For greater accessibility to non-specialists, the published versions include an extended introduction, as well as a foreword by the student's supervisor explaining the special relevance of the work for the field. As a whole, the series will provide a valuable resource both for newcomers to the research fields described, and for other scientists seeking detailed background information on special questions. Finally, it provides an accredited documentation of the valuable contributions made by today's younger generation of scientists.

Theses are accepted into the series by invited nomination only and must fulfill all of the following criteria

- They must be written in good English.
- The topic should fall within the confines of Chemistry, Physics, Earth Sciences, Engineering and related interdisciplinary fields such as Materials, Nanoscience, Chemical Engineering, Complex Systems and Biophysics.
- The work reported in the thesis must represent a significant scientific advance.
- If the thesis includes previously published material, permission to reproduce this must be gained from the respective copyright holder.
- They must have been examined and passed during the 12 months prior to nomination.
- Each thesis should include a foreword by the supervisor outlining the significance of its content.
- The theses should have a clearly defined structure including an introduction accessible to scientists not expert in that particular field.

K. Felix Mackenroth

Quantum Radiation in Ultra-Intense Laser Pulses

Doctoral Thesis accepted by
Ruperto-Carola University of Heidelberg, Germany

Springer

Author
Dr. K. Felix Mackenroth
Max Planck Institute for Nuclear Physics
Heidelberg
Germany

Supervisor
PD Dr. Antonino Di Piazza
Max Planck Institute for Nuclear Physics
Heidelberg
Germany

ISSN 2190-5053 ISSN 2190-5061 (electronic)
ISBN 978-3-319-07739-0 ISBN 978-3-319-07740-6 (eBook)
DOI 10.1007/978-3-319-07740-6
Springer Cham Heidelberg New York Dordrecht London

Library of Congress Control Number: 2014941089

Printed on acid-free paper

Springer is part of Springer Science+Business Media (www.springer.com)

To Sibylle, Geert, Hanna, Julia and Luisa

Supervisor's Foreword

The understanding of the emission of electromagnetic radiation lies at the very heart of understanding quantum electrodynamics. Emission processes from a charge carrier can efficiently be triggered by the interaction with accelerating electromagnetic fields. Consequently, there has been an ever-increasing interest in the emission from elementary particles subject to most violent accelerations in order to test the theory under most extreme conditions. A paramount example of such strong accelerations is delivered by high intensity lasers, providing unprecedentedly high electromagnetic field strengths ever since their invention. Analyzing the radiation resulting from this type of interaction thus is a prime means for pushing the understanding of quantum electrodynamics to its edge. Whereas the experimental evolution of available laser systems has been constantly proceeding, the theoretical description of the processes under consideration has for quite some while been relying on the assumption of the laser fields being monochromatic, worked out in the 1960s. This custom clearly failed to follow the current development of laser facilities toward ever shorter pulse durations yielding ever higher peak intensities. Moreover, there are several conceptional problems arising in the analysis of monochromatic laser fields such as infinitely narrow spectral lines or even divergences of the electron propagator. Furthermore, it meant the disregard of the laser's temporal evolution and sub-envelope dynamics via which, on the other hand, there are disclosed a number of possible further applications. Among these novel prospects, theoretically tractable only by properly taking into account the temporal evolution of the laser field, are effects of the laser's absolute phase or the emission of a number of photons larger than one. The present thesis provides insights into the outlined direction. Its numerical and analytical methods have been applied to the quantum electrodynamical theory of radiation in laser pulses of arbitrary temporal duration comprising down to only a few laser cycles. The calculations are carried out in the so-called scheme of strong-field quantum electrodynamics, taking a laser field of arbitrary strength into account exactly. In order to follow the natural cascade of complication this thesis is started with the investigation of the emission of only one single photon by an electron scattered from an intense laser pulse. The results of this analysis provide the first realistic proposal for determining the carrier envelope phase of ultrarelativistic and ultra short laser pulses. The next natural step then is to study the emission of more than one photon. Consequently, the coherent emission of two photons is studied, representing a higher order quantum electrodynamical process, implying that this latter process will inevitably be only a small correction to the dominating single photon emission and thus hardly detectable. It is demonstrated how this difficulty can be circumvented by properly understanding the angular distribution of the emitted radiation. The proposed detection scheme opens a door to the observation of higher order quantum electrodynamics effects in a yet untested regime. In particular, it links to a regime, where the electron dynamics becomes strongly affected by the radiation emitted by the electron itself, an effect referred to as radiation reaction. The present thesis thus sheds light on a full theory of electromagnetic radiation including several unconventional phenomena and the reader is invited to enjoy the tour to the very core of electrodynamics.

Heidelberg, April 2014 PD Dr. Antonino Di Piazza

Abstract

The theoretical framework for describing the emission by free charged particles, scattered from highly intense laser fields, is extended to arbitrary temporal shapes of the scattering laser field. This work is motivated by the recent trend of laser technology, to achieve highest laser intensities by a tight temporal compression of the laser energy, down to only a few cycles of the carrying electromagnetic wave. Since modern laser fields are inaccessible to the perturbative treatment of usual QED, they are described as unquantized external fields and taken into account exactly. The emission of one or two photons is particularly studied. For both processes a powerful analytical approximation is formulated, valid in the experimentally relevant regime of high laser intensities and electron energies. This technique foreshadows possible applications, such as a viable way of determining the absolute phase of a highly intense few-cycle laser pulses, which was an unresolved problem so far. Furthermore, it is demonstrated how the usually strongly suppressed signal from two photon emission can be reliably discriminated from the dominant single photon emission signal. Finally analytical solutions for two hitherto unresolved issues are presented: Describing the spatial focusing of a few-cycle laser pulse and solving the Dirac equation in the presence of a focused laser beam.

Acknowledgments

I feel greatly indebted to a large number of people, to whom I wish to express my deepest thanks to with these opening lines.

The greatest thanks I owe and want to express to my family members Sibylle, Geert, Hanna, Julia, and Luisa, who by their ceaseless support and encouragement made my studies as well as my whole life as I led up to now possible. You are at the same time the safest harbor and the strongest drive any person could possibly wish for. I sure hope for many happy returns of the great times we have whenever and wherever the family gathers.

Another person without whose constant help and encouragement this thesis would not exist, is my supervisor PD Antonino Di Piazza, to whom I am deeply indebted. I consider myself lucky to have in his person a scientific teacher whose profound knowledge and intuitive understanding of physics is a perpetual motivation and inspiration for my work. And a special thank you is in order for his careful proofreading of this thesis, which resulted in many pieces of good advice concerning style and content.

To Prof. Christoph H. Keitel I am grateful for harboring me in his theory division at the Max-Planck Institute for Nuclear Physics and providing an excellent research environment.

I thank Prof. Dirk Dubbers for agreeing to write the second review for my thesis. I also thank all members of my final examination committee: Antonino Di Piazza, Dirk Dubbers, Georg Wolschin and Werner Aeschbach-Hertig for agreeing to be on this committee as well as inspiring discussions about physics.

I acknowledge fruitful and inspiring discussions with Daniel Seipt.

To many members of the staff of the Max-Planck Institute for Nuclear Physics I also owe my thanks for providing a great deal of help in administrational and technical issues. In particular I thank Sibel Babacan, Peter Brunner and Dominik Hertel.

I am much obliged to all the truly helpful people who proofread this thesis: Stefano Cavaletto, Sebastian Meuren, Norman Neitz, Juliane Thalheim, and Anton Wöllert.

I thank my former office members Sebastian Meuren, Sarah Müller, and Benedikt Wundt for maintaining an as well entertaining as productive working atmosphere. The same holds for my present officemates Sven Ahrens, Enderalp Yakaboylu, and Wen-Te Liao.

I thank many members of the theory division of the Max-Planck Institute for Nuclear Physics, former and present: Adriana Pálffy, Michael Ruggenthaler, Matthias Ruf, Sarah Müller, Norman Neitz, Stefano Cavaletto: You guys seasoned the years at the institute with a good measure of fun at various occasions.

Particular gratitude I express to Jan Walkling and Patrick Zessin, who fairly often pulled me out of too deeply sophisticated thoughts and showed me the value of friendship.

Finally I thank all the people involved in the Springer Theses program for the opportunity of publishing this work. In particular I thank Dr. Angela Lahee for her interest in this thesis alongside several helpful suggestions for improvement and Mr. Anand Jayaprakash for his patience and help in formal matters.

Contents

Chapter 1
Introduction

> *Fiat lux.*
>
> Genesis 1:3
>
> *We can scarcely avoid the inference that light consists in the transverse undulations of the same medium which is the cause of electric and magnetic phenomena.*
>
> James C. Maxwell

The world as perceived by scientists is nowadays believed to be governed by four fundamental forces. The everyday world as accessible to our senses, dominated by the interaction of electrons in separate atomic shells, however, is largely steered by the electromagnetic force only [1]. It is thus plausible that the electromagnetic interaction is most thoroughly investigated to develop the best possible understanding of its phenomena. Despite the fact that most everyday phenomena are well described in the framework of classical electrodynamics (CED), it is commonly known that the superordinate theory is the theory of quantum electrodynamics (QED) [2]. The fundamental difference between the classical description of any type of radiation and the quantum concept is depicted in Fig. 1.1, visualizing a physical system capable of emitting radiation, such as a particle with momentum $\boldsymbol{p}$, entering an interaction region, where by some mechanism it is led to the emission of radiation. In CED the emitted radiation is described as a continuous distribution of an electromagnetic field, created by the radiating physical system (Fig. 1.1a). In quantum physics, on the other hand, the emitted radiation, such as any electromagnetic field, is composed of individual photons, whose entirety constitutes the quantum radiation (Fig. 1.1b). Since QED includes CED, the following discussion will be focussed on the former. There has been a plethora of tests for this theory and within the grasp of our current technological limits it exhibits unprecedented predictive power and has passed a variety of experimental test. It is then understandable that any disagreement between experimental measurements and theoretical predictions of QED, as, e.g., the just recently reported deviation of a measurement of the proton radius from the QED result by seven standard deviations [3], motivates a vast scientific debate. On the other hand, as an example of highly accurate QED predictions, think of the electron's g-factor,

K. F. Mackenroth, *Quantum Radiation in Ultra-Intense Laser Pulses*, Springer Theses, DOI: 10.1007/978-3-319-07740-6_1,

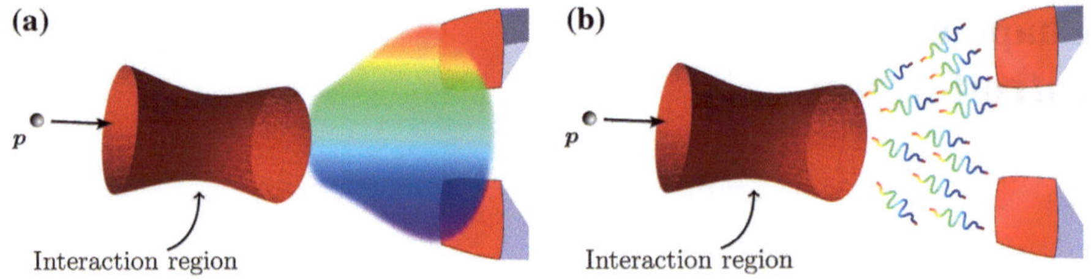

Fig. 1.1 Fundamentally different concepts of radiation emitted by a particle as viewed in classical (**a**) and quantum (**b**) physics

either in a free state [4] or bound to several different types of hydrogen-like atoms [5–7] (For a broader overview see [8] and references therein). One fundamental problem of QED, faced in any such calculation, however, is that it is a theory which is not solvable exactly (see Sect. 2.2). The generic way of obtaining meaningful results is expand the general result into a perturbation series in the small parameter e, the unit charge. To this day it is a highly nontrivial task of QED to obtain higher order terms of this perturbative series. Returning to the previously mentioned high-precision calculations of the electron's g-factor, this means that one has to compute various kinds of correction terms [9] such as relativistic mass corrections, nuclear shape effects [10], nuclear recoil [11–13] and QED loop corrections [11, 14–18]. Comparing the resulting theoretical values to experimental data, presently available for various species of highly charged ions such as $^{28}Si^{13+}$ [7], $^{16}O^{7+}$ [6] and hydrogen-like carbon [5], provides a generic test of QED as well as benchmark results for fundamental constants entering the theory. The tremendously good agreement between these calculations and experiments deemed the term of QED being the "best tested theory we have". The according conclusion that QED is a theory which is well investigated and understood, is, however, incomplete. In contrast, it is accepted that QED is not the final theory to correctly describe the electromagnetic interaction realized in our universe, owing to the cumbersome properties QED exhibits, such as infinite particle masses and charges, which are treated consistently in the framework of renormalization [19–21]. Because such divergences, however, emerge at tremendously high energies, where the testability of the theory is pushed to its limits, since the formulation of QED itself research was driven by the urge to explore such limits. One viable way of investigating these limits is to test QED at small scales, which are translated to high interaction energies. The highly precise g-factor calculations, mentioned above, are counted among the tests of this sector of QED. Another approach, complementary to this high-precision sector of QED, is realized by testing it in the presence of strong electromagnetic fields to push the theory to its limits.

In this respect, the significance of understanding a charged elementary particle's dynamics in the presence of a strong electromagnetic field was pointed out even before the formulation of QED. In 1929 Klein discovered a seeming contradiction of the single-particle quantum theory of the electron [22], which Dirac had presented a year earlier [23]: He noted that, according to Dirac's theory, the wave function of an electron with energy ε, scattered at a potential step of height V, for very large values of V is not necessarily completely reflected, in obvious contradiction to classical mechanics. (Just imagine a ball thrown at a wall would penetrate it if only it was thick enough!) In fact, Klein found that for a potential height exceeding the sum of the electron's energy and its rest mass energy $\varepsilon_{\text{r.m.}} = mc^2$, where c is the speed of light, i.e. for $(V > \varepsilon + mc^2)$, the electron is transmitted with finite non-negligible probability and changes its kinetic energy from positive to negative values. This problem, nowadays known as *Klein's paradox*, was further elucidated in 1931 by Sauter [24, 25], who revisited Klein's analysis, assuming a finite steepness of the potential step. Since when realizing an electrostatic potential step by a constant electric field this field is proportional to the gradient of the potential step, Sauter was able to investigate fields of finite strength, in contrast to the infinitely strong constant electric fields Klein had considered. He could show that the transmission coefficient could only be non-negligible, if the potential rise provided the electron with an energy comparable to its rest mass over its Compton wavelength. Realizing such a potential gradient by a constant electric field would require a field of the *critical field* strength

$$\mathcal{E}_{\text{cr}} = \frac{m^2 c^3}{\hbar |e|} \approx 1.3 \times 10^{16} \frac{\text{V}}{\text{cm}}, \tag{1.1}$$

where $e < 0$ is the electron's charge. Still today all controllable artificial electric fields available in laboratories around the globe fall short of reaching this field strength by several orders of magnitude. The same field threshold was obtained by Heisenberg and Euler in an analysis of spontaneous pair creation by an electric field in vacuum [26]. It was also these authors who coined the term *critical field*. They found the probability of an electric field of amplitude $\mathcal{E}$ to create an electron-positron pair to scale as $\exp\left(-\pi\mathcal{E}_{\text{cr}}/\mathcal{E}\right)$. In other words a constant electric field of amplitude $\mathcal{E}_{\text{cr}}$, corresponding to the critical potential gradient found by Sauter [25], spontaneously creates electron-positron pairs from vacuum. For electric fields far below the critical field $\mathcal{E} \ll \mathcal{E}_{\text{cr}}$ the probability is exponentially suppressed. An illuminating interpretation of Klein's paradox and its connection to pair creation from vacuum was given by one of the founding fathers of QED, Julian Schwinger, who put it into a field-theoretical framework [27]. He explained the process of pair creation from vacuum, steered by an electric field, as the coupling of the field to the virtual particle pairs of the QED vacuum, providing some pairs with sufficient energy to emerge into physical reality. This effect is often named after Schwinger, who first calculated the lowest order pair creation probability in a constant electric field, employing a QED framework. He found the probability of pair creation to be exponentially suppressed below a critical field strength, equivalent to the value found by Sauter and by Heisenberg and Euler in their single-particle analysis. The resulting habit, to frequently label this critical field *Schwinger field*, will not be followed in this thesis, for the outlined reasons. We will much rather stick to the term critical field (of QED), with the wording in brackets omitted when there is no confusion possible. The consequently formed strong-field sector of QED (or short: strong-field QED (SF-QED)) has since then proven a formidable testing ground for the frontiers of QED and is subject of an ever growing research community [28, 29]. Also from the above discussion one concludes that, albeit QED is valid for arbitrary physical systems, which underlie the electromagnetic interaction and are stable on a time scale much longer than the electromagnetic interaction time, its key aspects can be inferred from the analysis of electron dynamics. It is thus customary to refer to electrons as the involved massive particles, keeping in mind the broader applicability of the presented methods. One possible way to obtain experimental data from the interaction of electrons with intense electromagnetic fields is to detect the radiation emitted by the electron during the interaction. Investigating the properties of this radiation then serves as an excellent testing ground for the foundations of SF-QED itself as well as of other fundamental physics. Beyond that, it may even open access to physical processes which are otherwise not accessible to direct measurement. A prominent example of such a research field, which is untestable other than by means of ultra-strong electromagnetic fields, thus requiring a SF-QED description, is astrophysics [30]. Experimental strong-field setups are largely capable of providing, in earth-bound settings, the extreme temperatures, pressures or electromagnetic fields of stellar and cosmological objects, that are not accessible to direct experimental examination [31]. Since in a majority of astrophysical environments matter is considered as almost completely ionized, the radiation emitted in such extreme experimental setups is largely dominated by free electrons interacting with strong electromagnetic fields [32].

And the grasp of QED calculations even goes beyond this. In addition to so-called *classical effects*, which feature an analog in CED, QED predicts new, fully non-classical effects. A prime example of this class of effects is the simultaneous emission of two photons by a free electron. This so-called double Compton scattering was investigated in the perturbative regime of ordinary QED just after its formulation itself [33] and experimentally confirmed shortly afterwards [34]. Also this class of QED effect, which do not feature a classical limit, was carried over to the SF-QED regime. Quite recently explicit calculations were presented, analyzing the probability of the simultaneous emission of two photons in a strong electromagnetic field [35, 36]. Such calculations may eventually help to resolve an ongoing fundamental discussion on the so-called *Unruh effect*: This notion refers to the effect that an electron experiencing an acceleration in a vacuum would not experience its surroundings as the vacuum state but much rather as a thermal bath. Any interaction with the thermal bath, experienced by the accelerated electron is believed to be translatable to the emission of two photons [37–39], which can be rigorously described only in the framework of QED. There have been several experimental schemes proposed as to how the Unruh effect can be detected by means of strong-field experimental schemes [40–42].

We have thus seen a cornucopia of facets of the interaction of intense electromagnetic fields with free electrons, whose understanding consequently is of particular interest. The possibilities of generating the necessary highly intense electromagnetic fields in a laboratory environment are limited to two conceptual cases, to which consequently also all theoretical work is restricted. A first possible way is given by the investigation of electric fields of highly charged atomic nuclei acting on the electrons in the atomic shell [43]. Applications of this branch of physics are fundamental QED

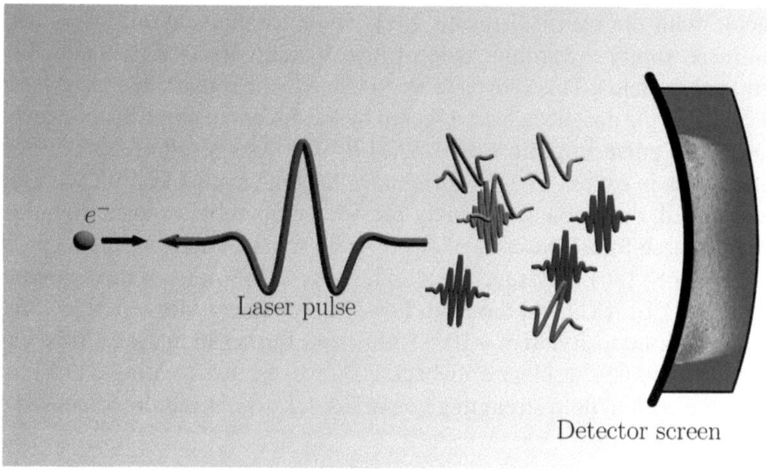

Fig. 1.2 Conceptual picture of an electron scattered from a strong laser pulse and emitting radiation

processes occurring in the collision of highly charged ions, such as pair creation [44]. Furthermore, the theoretical description of such fields is employed to stimulate the above mentioned high-precision sector of QED, by providing additional correction terms to the g-factor [10]. Another route to the generation of strong electromagnetic fields has been opened up by the invention of the laser [45], as pictured in Fig. 1.2, which differs from the original Fig. 1.1 in as much as here the radiating system is fixed to be an electron and the excitation, stimulating the electron to radiate, is caused by a laser field. In fact, the development of laser technology has widely steered the theoretical works of significant parts of SF-QED research, since the electromagnetic fields generated by a laser are notedly well controllable and allow a particularly clear interpretation of experimental measurements. As soon as SF-QED was formulated, its methods were utilized to analyze the interaction of two laser waves, mediated by the virtual electron-positron pairs of the QED vacuum [46–49], the emission of a single photon from an electron scattered in a laser field [50–61], pair creation in a laser field [51, 61–63] and atomic ionization by absorption of many photons from the laser field (above threshold ionization (ATI)) [64–66]. Furthermore, fundamental quantities such as the electron's Green's function in the presence of a strong laser wave, also called *dressed propagator*, were investigated [61, 67, 68]. Interest in these topics has not diluted even until today, constantly encouraging the upcoming of analyses for previously undocumented physical phenomena (see [29] for an extensive overview over the voluminous work in the field). Alongside this trend of studying QED effects, due to the emergence of the laser, providing unprecedented electric field strengths, there emerged a strong interest in describing electron dynamics in highly intense laser waves within the framework of CED [69–72], which has also prevailed until today [73–75]. The quantum and classical analyses were multiply interwoven with each other, as the classical limits of many quantum computations were explicitly shown to coincide with the classical results [61]. Since their invention, lasers have thus been an increasingly rewarding route of investigating electron dynamics in strong electromagnetic fields. This is even more true due to the remarkable progress in laser technology over the past decades, triggered by the advent of amplification techniques such as chirped pulse amplification (CPA) [76–78]. As a result of this development, laser intensities in excess of 10^{22} W/cm² have become available [79], with facilities being envisaged, designed to routinely provide comparable or even higher intensities. Among these future beacons of strong-field science there are the petawatt field synthesizer (PFS) [80] (envisaged peak intensity: 10^{22} W/cm²) or the Extreme Light Infrastructure (ELI) [81] and the High Power Laser Energy Research [82] facilities (envisaged peak intensity: $10^{25}-10^{26}$ W/cm²) and further increase is to be expected for the future [29, 83]. A clear drawback is, however, that creating an electric field exhibiting the critical field strength given in Eq. 1.1, would require a focused optical laser peak intensity of

$$I_{\text{cr}} = \frac{c}{4\pi}\mathcal{E}_{\text{cr}}^2 = 4.6 \times 10^{29} \frac{\text{W}}{\text{cm}^2}. \tag{1.2}$$

Until today all laser facilities, and even those planned, underrun this critical value by 3–4 orders of magnitude. It is, however, not the peak laser intensity, available in a laboratory, which is the important parameter characterizing the interaction of a laser field with a relativistic electron: The electron will necessarily respond to the electric field it experiences in its rest frame, but there the laser's electric field, can be significantly larger than the one measured in the laboratory, due to the Lorentz boost. The decisive parameter thus is a Lorentz invariant parameter, which measures the electric field in the electron's rest frame, which is labeled χ and defined below. Strong-field facilities around the globe consequently provide an unequaled testing ground for the basic principles of SF-QED and even QED itself. Therefore, the scientific prospects given by strong-field science are highly promising. In fact, it was shown that already in laser fields of readily available intensities new physical regimes are reached. For instance, electrons exposed to a laser field exceeding an intensity of 10^{18} W/cm² experience violent accelerations to relativistic speeds within less than the field's oscillation period. This regime has consequently been dubbed *relativistic optics* and lasers reaching such tremendous intensities are called *relativistically intense* [28, 84]. It has proven useful to quantify the interaction strength of a relativistically intense laser pulse, featuring a peak electric field strength $\mathcal{E}_L$ and a frequency spectrum centered at $\nu_L = \omega_L/2\pi$, with an electron in terms of the Lorentz invariant parameter [50]

$$\xi = \frac{|e|\,\mathcal{E}_L}{\omega_L\, m\, c}. \tag{1.3}$$

In this expression the subscript L refers to *Laser*. The parameter ξ can now, since it does not contain the Planck constant $\hbar$, be physically interpreted as a classical parameter in the sense that also classical computations ($\hbar \to 0$) will be sensitive to ξ.

To develop an intuitive understanding of the physical significance of this parameter, one may employ the classical picture of a point-like particle being scattered off a uniform electromagnetic field. A classical electron, accelerated by a constant electric field of strength $\mathcal{E}_L$ over a timespan τ, reaches a velocity of

$$\beta = \frac{|e|\,\mathcal{E}_L}{mc}\tau. \tag{1.4}$$

Having in mind that the characteristic timespan of an electromagnetic field oscillating at the angular frequency ω_L is its cycle period $\tau \sim \omega_L^{-1}$, we conclude that the parameter ξ can be interpreted as the fraction of the speed of light an electron is approximately accelerated to within one laser period [61, 85]. For the interaction of an electron with an optical laser pulse ($\hbar\omega_L \sim 1$ eV) this parameter reaches unity in electric fields of the strength $\mathcal{E}_L^{\text{opt}}(\xi \sim 1) \sim 10^{10}$ V/cm. To realize such an electric field by a laser requires intensities of the order of

$$I_L^{\text{opt}}\,(\xi \sim 1) \sim 10^{18}\frac{\text{W}}{\text{cm}^2}, \tag{1.5}$$

which is equal to the aforementioned threshold of the relativistic regime. This regime is now understood to be distinguished by $\xi \gtrsim 1$. We stress that laser fields of this intensity are routinely available nowadays [28] and the indicated realm of ξ exceeding unity can be thoroughly studied. The record optical intensity of 2×10^{22} W/cm^2, reported in 2008 [79], even corresponds to a parameter $\xi \sim 10^2$. Due to the frequency dependence of ξ, for X-ray frequencies ($\hbar\omega_L \sim 1$ keV) this translates to $\mathcal{E}^{\text{X-ray}}\,(\xi \sim 1) \sim 10^{13}$ V/cm and a corresponding laser intensity of $I^{\text{X-ray}}\,(\xi \sim 1) \sim 10^{24}$ W/cm^2.

Also on a formal level, the parameter ξ plays a decisive role: As was mentioned above ordinary QED computations are only possible in terms of a perturbation series with the expansion parameter the electric charge e. It was now shown that in the presence of a very dense photon field, as is provided by an intense laser, the expansion parameter for higher orders of photon absorption from the laser field on the amplitude level no longer is e, but the parameter ξ [50, 61]. The probability of absorbing n photons from the laser field mode accordingly scales $P^n \sim \xi^{2n}$. For relativistically strong laser fields, but still fulfilling $\xi \lesssim 1$, one then has to sum over a large number of interaction orders. An even worse behaviour emerges as soon as ξ grows larger than unity: The expansion parameter of the perturbation series in the interaction with the strong external laser field is larger than unity and the whole perturbation theory breaks down. Therefore, ordinary QED cannot be deployed as a theoretical tool in this case, for reasons of either impracticality or even inapplicability. The accordingly necessary alternative approach to treat strong external fields is given by the formal foundations of SF-QED, whose physical aspects were discussed above. This formal framework, albeit widely investigated and understood, is far from being complete. Within the plethora of new phenomena, as predicted by SF-QED, there are still many not properly understood or experimentally not yet observed, such as the effect of mass dressing [70, 86, 87] or the intensity dependent red shift in nonlinear Compton scattering (which essentially amount to the same effect) [55, 88] and possibly many completely undescribed and unexpected phenomena. Due to lack of sufficiently intense laser systems so far, SF-QED effects have been observed in only a limited number of experiments [89], as there are nonlinear Compton scattering [90–92] and pair production [93]. With the development of laser systems to ever higher peak intensities, more experimental tests seem to be in reach.

The parameter ξ is not the only Lorentz invariant parameter that may be constructed from the four vectors entering the problem of a free electron of momentum p^μ interacting with a laser pulse. In principle one can construct several further invariant parameters, characterizing the scattering, but in the important case of the laser being modeled by a plane wave field with wave vector k_L^μ and four potential A_L^μ there is only one further non-vanishing invariant, which is defined as [61]

$$\chi = \frac{|e|\hbar\sqrt{|\left(F_L^{\mu\nu}p_\nu\right)^2|}}{c^3m^3} = \frac{(pk_L)}{mc\,\omega_L}\frac{\mathcal{E}_L}{\mathcal{E}_{\text{cr}}}. \tag{1.6}$$

$F_L^{\mu\nu}$ is the electromagnetic field strength tensor (see Sect. 2.1) and the critical field is given in Eq. (1.1). In a reference frame where the electron initially is at rest ($p^\mu = (mc, \mathbf{0})$), this parameter reduces to $\chi = \mathcal{E}_{\text{r.f.}}/\mathcal{E}_{\text{cr}}$, where the subscript refers to *rest frame*. The parameter χ consequently quantifies the laser's electric field in units of $\mathcal{E}_{\text{cr}}$, as an electron experiences it in its rest frame. In contrast to ξ this parameter contains the reduced Planck constant $\hbar$, whence it is connected to quantum effects. The radiation of an electron, scattered from an intense laser wave, is influenced by two classes of such quantum effects, namely the quantization of its motion and the recoil it experiences upon emitting a photon with an energy comparable to the electron's instantaneous energy. The first class of quantum effects is due to the non-commutativity of the electron's position and momentum, which was shown to be negligible for a highly relativistic particle [94]. The notion *quantum effects* in analyzing photon emission from a highly relativistic electron, thus solely refers to photon recoil. This observation translates to the notion that the motion of the electron in a plane wave field is quasi-classical and its dynamics can be described in terms of a classical trajectory, except specifically the photon emission. Inside a plane wave field the recoil an electron experiences upon emission of a photon with wave vector k'^μ becomes significant when the following condition is met

$$\frac{k'k_L}{pk_L} \sim 1. \tag{1.7}$$

In the regime $\xi \gg 1$ this condition corresponds to $\chi \sim 1$ and, accordingly, one expects recoil to become significant in this regime [61]. This corresponds to a scaling of the recoil with the laser's electric field strength and is confirmed by experimental data [93]. This is a clear deviation of the behaviour in the linear regime, where it was a milestone experiment of the early 20th century to find that the recoil only scales with the photon energy [95].

From the presented considerations it follows that the interaction of electrons with highly intense laser fields is a vivid area of research and rightfully claims a prominent position in the investigation of the foundations of QED. We furthermore found that the rapidly developing laser technology is well suited to provide researchers with the desired strong electromagnetic fields. The remarkable interest in the interaction of electrons with laser fields, however, is not only motivated by fundamental physical questions as outlined above. Employing lasers of the mentioned unprecedented powers has additionally opened up access to nowadays widely used technical applications such as small-scale laser-based particle accelerators [96–99], which are anticipated to be even suitable for medical applications [100], or novel technical challenges, such as laser ignited nuclear fusion [101, 102]. Even more, the radiation emitted by electrons during the interaction with an intense laser beam may prove to be of tangible technical meaning. For instance, an electron with momentum p^{μ}, colliding with a photon with wave vector k_L^{μ} will emit a photon with a wave vector $k'^{\mu} = \hbar\omega' n'^{\mu}$, featuring a Doppler-shifted energy of

$$\hbar\omega' = \frac{pk_L}{n'p}. \tag{1.8}$$

When a relativistic electron ($\varepsilon \gg m$) collides head-on with a beam of photons with energy $\hbar\omega_L$, then for the photons backscattered into the electron's direction of propagation Eq. (1.8) implies a boost of the backscattered photons' frequency, according to $\omega' = 4\,(\varepsilon/m)^2\,\omega_L$ [103–105]. The corresponding tremendous upconversion factors can be employed to efficiently generate high-energy photon beams, for which there exists a plethora of possible applications. Apart from this practical use for technological progress, lasers are demonstrated to serve as highly utile tools for fundamental research such as nuclear and particle physics [106–108], radiation reaction [109], or for elucidating properties of the vacuum of quantum field theory, such as its role as mediator of photon-photon interaction [110–113]. We accordingly find it highly desirable to develop a profound understanding of the interaction of a free electron with an intense laser field to fully understand the vast prospects of strong-field science. Interestingly, it is also found a highly beneficial influence of this research on the development of laser technology. Since the energy densities of available laser systems underwent the described immense increase, the established schemes for characterizing laser fields are no longer applicable to them, mainly due to the radiation damage inflicted by the ultra-strong laser flashes. In contrast to this lack of measurability, it was found that, by virtue of theoretical methods developed in the past few years, the understanding of SF-QED may pave the road to new schemes to thoroughly characterize ultra-intense laser fields. There have been experimental schemes proposed to measure laser intensities based on atomic ionization [114] or from the angular distribution of the radiation emitted by an electron scattered from the laser pulse under investigation [115]. Furthermore it was shown that even the precise field shape of an ultra-intense laser pulse can be inferred from the radiation emitted by an electron scattered from the pulse [116]. We thus find the enrichment of laser physics and SF-QED to be mutual. Furthermore it was demonstrated how the degree of entanglement of two photons, emitted simultaneously by an electron scattered from a strong laser wave, could be quantified [35, 36]. The entanglement of two separate quantum objects, such as two spatially separable photons, is a key ingredient for quantum information and quantum cryptography [117, 118]. Though the technical applications in these fields are usually realized at only moderate photon energies, the understanding and control of correlated high-energy photon pairs may open up a new access to these rapidly developing branches of physics.

A common feature of all the laser facilities mentioned above is the road taken for achieving the tremendous laser powers. The already mentioned chirped pulse amplification, which is a working horse of laser systems reaching peak powers in the PW regime ($P > 10^{15}$ W), in simple words is comprised of three steps. The laser pulse, that is to be amplified, is stretched to a duration considerably longer than it features initially, whence its intensity is reduced, thus preventing the optical components from physical damage. The stretched pulse is then amplified in a conventional laser amplifier and this amplified pulse is compressed back to a very short duration [83]. Already from this basic idea it is clear that a key ingredient to obtain high laser powers are shortest pulse durations. The highest laser intensities are naturally achieved by compressing the pulses to the threshold of the lowest possible durations. That there is such a lower threshold is obvious from the consideration that only oscillating functions can be propagating solutions of the Maxwell equations. Thus, in principle at least one oscillation of the laser wave has to be contained in the pulse. We do not discuss the commonly applicable generation of half-cycle pulses, which, however, is possible only by elaborate use of emission characteristics of specific media [119]. The lower threshold for laser pulse durations is then laid by the fundamental oscillation period of the laser field ω_L^{-1}. Hence, a high-power laser pulse should comprise only very few cycles of the carrying electromagnetic field, whence the term *few-cycle pulses* was coined. Such pulses are distinguished by their duration τ_L fulfilling the condition

$$\omega_L \tau_L \sim 1. \tag{1.9}$$

Such pulses have become available in a wide range of frequencies such as in the mid-infrared ($\tau_L \approx 39$ fs, $\hbar\omega_L = 0.10$ eV) [120], in the near-infrared ($\tau \approx 4.3$ fs, $\hbar\omega_L = 0.8$ eV) [121], in the optical ($\tau < 4$ fs, $\hbar\omega_L = 1.77$ eV) [122], and in the extreme ultraviolet regimes ($\tau \approx 80$ as, $\hbar\omega_L = 103$ eV) [123, 124] and almost all future planned strong-field laser facilities are designed to operate in the few-cycle regime. In contrast to this technical development there stands the state of affairs in theoretical SF-QED. Almost all of the above cited early works on the interaction of electrons with intense laser fields were obtained for monochromatic laser waves. Due to their simple spectrum, i.e. a δ-spike in frequency space, they were investigated early on and analytically well understood (see Sect. 2.2.3). Due to the infinite temporal extent of the scattering field, however, it was only possible to obtain time averaged quantities such as the cross section, where much information about the scattering field itself is lost. Such a description is clearly disfavorable in light of the general route of temporal and spatial pulse compression. Even though there has been some early work on non-monochromatic laser pulses [125], in this work the authors assume the temporal focusing of the laser pulse to be only mildly and hence subject their analysis to the additional condition $\omega_L \tau_L \gg 1$. Albeit this assumption is still well justified for presently available optical high-power laser facilities, which operate at pulse durations of several tens of fs, it is not suited to follow the present trend of high-intensity laser physics to employ only few-cycle laser pulses. The cited work thus is capable of describing presently used many-cycle laser pulses, whose emission spectra largely resemble the monochromatic case. On the other hand, key aspects of the properties of few-cycle laser pulses, as will be abundantly used in the near future, are not captured.

This thesis is meant to contribute to closing this gap. We are going to investigate radiation emitted by an electron upon interaction with an intense laser pulse of arbitrary temporal shape, following the technical trends and foremost frontiers of strong-field research. Emission processes in which the electron can change its momentum are known as Compton scattering, in analogy to the original Compton effect [95]. One additionally labels the scattering as nonlinear, since the process may exhibit a nonlinear dependence on the intense laser's field strength. In the remainder of this introductory chapter we will introduce typical notation, frequently used in this work and light-cone coordinates. These latter are a set of coordinates, particularly fit and favorable for the investigation of QED scattering matrix amplitudes for the interaction with plane wave fields.

In Chap. 2 we are going to present the fundamental methods of treating electron emission in CED as well as in SF-QED. We show how the well-known monochromatic analyses in CED and SF-QED give some equal, as well as some differing results. Furthermore, we present the canonical way of treating monochromatic focused laser beams in CED as well as its drawbacks when considering a spectrally broad laser pulse. We are going to demonstrate a possible way around this deficiency by treating the separate Fourier components of a laser pulse each as separately focused. Chapter 2 is concluded by a concise introduction to the SF-QED computation rules.

In Chap. 3 we turn to the simplest process of SF-QED, the emission of one single photon from an electron, so-called nonlinear single Compton scattering (NSCS). We are going to carefully work out a stationary phase approximation to the scattering matrix element, valid in the ultra-relativistic regime of high laser intensities and electron energies. Relying on this approximation, in Sect. 3.4 we present a possible way of inferring the exact shape of the scattering laser pulse from the angular region into which the electron emits radiation. We demonstrate that this determination scheme is quite robust and insensitive to numerous possible error sources, which we explicitly discuss. Finally we comment on an ongoing scientific discourse about the notion of a dressed mass of an electron scattered from an intense laser field. The limiting cases of a perturbatively weak and a monochromatic laser field as well as the classical case, where quantum effects are negligible, are be discussed in Appendix D.

In Chap. 4 we investigate the next higher order term of the perturbative series in the external field. In particular we study the emission of two photons from an electron scattered from a strong laser wave, so-called nonlinear double Compton scattering (NDCS). The stationary phase approximation for NDCS will be shown to be largely equivalent to that worked out for the case of NSCS. We will further analyze a common split-up of higher order SF-QED amplitudes into two partial amplitudes, which are commonly called *incoherent* and *coherent* contributions. To distinguish these partial amplitudes we demonstrate a natural splitting of the dressed electron propagator into two parts and discuss their physical interpretation. Finally, we are going to discuss how the two photon emission spectra become correlated in the emitted photons' energies and emission directions, as soon as photon recoil becomes significant. The perturbative and monochromatic limits are sketched in Appendix E.

Chapter 5 is dedicated to a short discussion of the Filon approximation. This approximation scheme for highly oscillatory integrals is the numerical work horse, employed for obtaining the exact spectra shown in this thesis. We discuss the approximation techniques for uni- and bivariate integrals separately. For the latter, however, we will make use of the fact that for the obtained quantum radiation matrix elements the integrands are almost fully separable in the two integration variables.

The concluding remarks of Chap. 6 then finalize the thesis.

1.1 Units and Notations

We begin by giving the natural constants which are needed for this work

Natural constants according to [126]	
$c = 2.998 \times 10^8 \mathrm{m\,s}^{-1}$	Speed of light
$\hbar = h/2\pi = 1.055 \times 10^{-34} \mathrm{J\,s}$	Planck's constant
$e = -1.602 \times 10^{-19} \mathrm{C}$	Electron charge
$\alpha_{QED} = (137.0)^{-1}$	Fine structure constant
$mc^2 = 511.0\,\mathrm{keV}$	Electron rest energy
$\varepsilon_0 = 8.854 \times 10^{-12}\,\mathrm{F\,m}^{-1}$	Vacuum permittivity

Whenever doing explicit calculations, however, we are going to use the natural unit conventions $\hbar = c = 1$. Other units are derived by the Gaussian units convention. Then the electromagnetic units are coupled to the base units via the relation

$$4\pi\varepsilon_0 = 1, \tag{1.10}$$

and the unit charge is connected with the fine structure constant via

$$e^2 = \alpha_{QED} \approx \frac{1}{137}. \tag{1.11}$$

In this work the charge of an elementary particle always contains its sign such that for an electron it holds $e = -|e|$. However, despite the Gaussian unit convention we employ for our calculations, at some points in this work characteristic quantities of experimental facilities or fundamental quantities are given in the standard SI units, for better comparability. Such statements are explicitly marked. Laser intensities can be derived from the parameters used in this work via the connections

$$1\frac{\mathrm{W}}{\mathrm{cm}^2} = 8.90037 \times 10^{17}\,\xi^2 \left(\frac{\omega_L}{\mathrm{eV}}\right)^2 = 2.32411 \times 10^{29} \left(\frac{\chi}{\gamma(1+\beta)}\right)^2. \tag{1.12}$$

Finally we give an overview over some useful conversion factors between natural units and their SI counterparts and important notation conventions used mostly without explicit explanation in this work

Unit conversion factors			
Distance:	$1\,(\mathrm{eV})^{-1}$	=	$1.9733 \times 10^{-7}\mathrm{m}$
Time:	$1\,(\mathrm{eV})^{-1}$	=	$6.5821 \times 10^{-16}\mathrm{s}$
Mass:	$1\,\mathrm{eV}$	=	$1.7827 \times 10^{-36}\mathrm{kg}$
Energy:	$1\,\mathrm{eV}$	=	$1.6022 \times 10^{-19}\mathrm{J}$
Electric field strength:	$1\,(\mathrm{eV})^2$	=	$4.3297 \times 10^{3}\mathrm{Vcm}^{-1}$
Power:	$1\,(\mathrm{eV})^2$	=	$2.4341 \times 10^{-4}\mathrm{W}$
Inverse area:	$1\,(\mathrm{eV})^2$	=	$2.5682 \times 10^{9}\mathrm{cm}^{-2}$

Mathematical conventions/notations	
$a^\mu = (a^0, \boldsymbol{a})$	Four vector, spatial components $\boldsymbol{a}$ in canonical coordinates
x^μ, p^μ	Four coordinates and electron four momentum
$u^\mu = p^\mu/m$	Four velocity
$\boldsymbol{\beta}$	Spatial particle velocity
$\gamma = \left(1 - \beta^2\right)^{-1/2}$	Lorentz factor of a particle with velocity $\boldsymbol{\beta}$
k_L^μ, ϵ_L^μ	Laser wave vector and polarization four vector
$f(x)$	Function depending on all space-time coordinates
$f(\boldsymbol{x})$	Function f depending on the spatial coordinates
$\partial_x f(x)$	Partial derivative with respect to x
$\partial_\mu = (\partial_t, \nabla)$	Four derivative
$a^\parallel$	Spatial component of a^μ parallel to $\boldsymbol{k}_L$
a^{pol}	Spatial component of a^μ parallel to $\boldsymbol{\epsilon}_L$
$\boldsymbol{a}^\perp = (a_1^\perp, a_2^\perp)$	Two spatial components of $\boldsymbol{a}$ perpendicular to $a^\parallel$
$a_\perp$	l.c.c. components of a^μ except a^-
$a^\perp$	l.c.c. components of a^μ except a^+
$g^{\mu\nu}$	Metric in canonical coordinates with signature $(+, -, -, -)$
$a_\mu b^\mu = (ab)$	Four product using Einstein's sum convention
γ^μ	Vector of the Dirac matrices
$\not{a} = \gamma_\mu a^\mu$	Feynman slash notation
$\hat{a}$	Quantum mechanical operator
$d^n a$	Volume element of the n arbitrary components of a
$d\boldsymbol{a}$	Volume element of the 3 spatial components of a vector (e.g. $\boldsymbol{x}$, $\boldsymbol{p}$) in canonical coordinates
$[\hat{a}, \hat{b}]_\pm = \hat{a}\hat{b} \pm \hat{b}\hat{a}$	(anti-) commutator of the operators $\hat{a}$, $\hat{b}$
$\mathbb{1}^n$	Unit matrix in n dimensions
l.c.c.	Light cone coordinates
m.c.	Monochromatic
L	Laser

1.2 Light-Cone Coordinates

Since a plane wave field defined by its wave vector k_L^μ depends on the four space-time coordinates only via the invariant combination $\eta = k_L^\mu x_\mu$, it is reasonable to employ a special set of coordinates where the plane wave field depends on only one coordinate instead of two, as in the case of canonical coordinates. Defining a unit vector parallel to the laser's propagation direction $\boldsymbol{n}_L = \boldsymbol{k}_L/\omega_L$, any vector $\boldsymbol{a}$ can be separated into a parallel $a^\parallel = \boldsymbol{a}\boldsymbol{n}_L$ and two perpendicular components, lying in the plane defined by $\boldsymbol{a}^\perp \boldsymbol{n}_L = 0$. We note that the components of the spatial wave vector of the plane wave in this reference frame is given by $k_L^\parallel = \omega_L$ and $\boldsymbol{k}_L^\perp = 0$. The resulting four dimensional coordinate system is called *light-cone coordinates* and is defined via

$$x^\mu_{\text{l.c.c.}} = \begin{pmatrix} x^- := \dfrac{t - x^\parallel}{\sqrt{2}} \\ x^+ := \dfrac{t + x^\parallel}{\sqrt{2}} \\ \boldsymbol{x}^\perp \end{pmatrix}. \tag{1.13}$$

To compute the invariant phase in these coordinates we need to compute a four vector product $\eta = k_L^\mu x_\mu = g_{\mu\nu} x^\nu k_L^\mu$ which requires the metric $g_{\mu\nu}$. This is easily computed to be

$$g_{\mu\nu}^{\text{l.c.c.}} = \frac{\partial x^\alpha}{\partial x^\mu_{\text{l.c.}}} \frac{\partial x^\beta}{\partial x^\nu_{\text{l.c.}}} g_{\alpha\beta} = \begin{pmatrix} 0 & 1 & 0 & 0 \\ 1 & 0 & 0 & 0 \\ 0 & 0 & -1 & 0 \\ 0 & 0 & 0 & -1 \end{pmatrix} = g^{\mu\nu}_{\text{l.c.c.}}. \tag{1.14}$$

We thus infer that the co- and contravariant vector components in l.c.c. are connected via

$$x^+ = x_-, \qquad x^- = x_+, \qquad \boldsymbol{x}_\perp = -\boldsymbol{x}^\perp,$$

whence some care is required on whether the l.c.c. index is a sub- or superscript. Thus in light-cone coordinates four products read

$$a_\mu b^\mu = a^+ b^- + a^- b^+ - \boldsymbol{a}^\perp \boldsymbol{b}^\perp \tag{1.15}$$

$$a^2 = 2a^+ a^- - \left(\boldsymbol{a}^\perp\right)^2 \tag{1.16}$$

The invariant phase is then simply given by $\eta = k_L^+ x^-$, since according to Eq. (1.13) all other components of k_L^μ vanish in l.c.c. This expression indeed depends only on one of the coordinates as was desired. Since we have to compute integrals over all four space-time coordinates several times we simply state the Jacobian of the coordinate transformation to l.c.c.

$$\left| \frac{\partial \left(x, x^{\parallel} \right)}{\partial \left(x^{+}, x^{-} \right)} \right| = 1. \tag{1.17}$$

The integration volumes over the spatially perpendicular coordinates is unchanged. We thus find for a differential four volume $dx^{-}dx^{+}d^{2}\boldsymbol{x}^{\perp} = dtdx^{\parallel}d^{2}\boldsymbol{x}^{\perp}$ which translates to finite volumes according to

$$\Delta x^{-} \Delta x^{+} = TL, \tag{1.18}$$

where $\Delta x^{\pm}$ is an arbitrary length in $x^{\pm}$-direction and T (L) is the finite length in t $(x^{\parallel})$-direction over which the equivalent four volume extends in canonical coordinates.

References

1. E.B. Goldstein, *Sensation and Perception*, 8th edn. (Wadsworth Publishing Company, Independence, KY, 2009)
2. Nobel Prize in Physics 1965. http://www.nobelprize.org/nobel_prizes/physics/laureates/1965/index.html
3. R. Pohl, A. Antognini, F. Nez, F.D. Amaro, F. Biraben, J.M.R. Cardoso, D.S. Covita, A. Dax, S. Dhawan, L.M.P. Fernandes, A. Giesen, T. Graf, T.W. Haensch, P. Indelicato, L. Julien, C. Kao, P. Knowles, E. Le Bigot, Y. Liu, J.A.M. Lopes, L. Ludhova, C.M.B. Monteiro, F. Mulhauser, T. Nebel, P. Rabinowitz, J.M.F. dos Santos, L. Schaller, K. Schuhmann, C. Schwob, D. Taqqu, J.F.C.A. Veloso, F. Kottmann, The size of the proton. Nature **466**, 213 (2010)
4. D. Hanneke, S. Fogwell, G. Gabrielse, New measurement of the electron magnetic moment and the fine structure constant. Phys. Rev. Lett. **100**, 120801 (2008)
5. H. Häffner, T. Beier, N. Hermanspahn, H. Kluge, W. Quint, S. Stahl, J. Verdú, G. Werth, High-accuracy measurement of the magnetic moment anomaly of the electron bound in hydrogenlike carbon. Phys. Rev. Lett. **85**, 5308 (2000)
6. J. Verdú, S. Djekić, S. Stahl, T. Valenzuela, M. Vogel, G. Werth, T. Beier, H. Kluge, W. Quint, Electronic g Factor of hydrogen like oxygen $^{16}O^{7+}$. Phys. Rev. Lett. **92**, 093002 (2004)
7. S. Sturm, A. Wagner, B. Schabinger, J. Zatorski, Z. Harman, W. Quint, G. Werth, C.H. Keitel, K. Blaum, g Factor of hydrogenlike $^{28}Si^{13+}$. Phys. Rev. Lett. **107**, 023002 (2011)
8. S.G. Karshenboim, Precision physics of simple atoms: QED tests, nuclear structure and fundamental constants. Phys. Rep. **422**, 1 (2005)
9. T. Beier, I. Lindgren, H. Persson, S. Salomonson, P. Sunnergren, H. Häffner, N. Hermanspahn, g_j Factor of an electron bound in a hydrogen like ion. Phys. Rev. A **62**, 032510 (2000)
10. J. Zatorski, N.S. Oreshkina, C.H. Keitel, Z. Harman, Nuclear shape effect on the g factor of hydrogen like ions. Phys. Rev. Lett. **108**, 063005 (2012)
11. T. Beier, The g_j factor of a bound electron and the hyperfine structure splitting in hydrogen like ions. Phys. Rep. **339**, 79 (2000)
12. V.M. Shabaev, V.A. Yerokhin, Recoil correction to the bound-electron g factor in H-like atoms to all orders in αZ. Phys. Rev. Lett. **88**, 091801 (2002)
13. K. Pachucki, Nuclear mass correction to the magnetic interaction of atomic systems. Phys. Rev. A **78**, 012504 (2008)
14. H. Grotch, Electron g factor in hydrogenic atoms. Phys. Rev. Lett. **24**, 39 (1970)
15. V.A. Yerokhin, P. Indelicato, V.M. Shabaev, Evaluation of the self-energy correction to the g factor of S states in H-like ions. Phys. Rev. A **69**, 052503 (2004)

16. K. Pachucki, U.D. Jentschura, V.A. Yerokhin, Nonrelativistic QED approach to the bound-electron g factor. Phys. Rev. Lett. **93**, 150401 (2004)
17. R.N. Lee, A.I. Milstein, I.S. Terekhov, S.G. Karshenboim, Virtual light-by-light scattering and the g factor of a bound electron. Phys. Rev. A **71**, 052501 (2005)
18. K. Pachucki, A. Czarnecki, U.D. Jentschura, V.A. Yerokhin, Complete two-loop correction to the bound-electron g factor. Phys. Rev. A **72**, 022108 (2005)
19. L.D. Landau, E.M. Lifschitz, *Lehrbuch der Theoretischen Physik - Band 4: Quantenelektrodynamik* (Akademie Verlag, Berlin, 1991)
20. M.E. Peskin, D.V. Schroeder, *An introduction to Quantum Field Theory* (Westview Press, Boulder, 1995)
21. W. Greiner, J. Reinhardt, *Quantum Electrodynamics* (Springer, Heidelberg, 2002)
22. O. Klein, Die reflexion von elektronen an einem potentialsprung nach der relativistischen dynamik von dirac. Z. Phys. **53**, 157 (1929)
23. P.A.M. Dirac, The quantum theory of the electron. P. Roy. Soc. Lond. A Mat. **117**, 610 (1928)
24. F. Sauter, Über das Verhalten eines Elektrons im homogenen elektrischen Feld nach der relativistischen Theorie Diracs. Z. Phys. **69**, 742 (1931)
25. F. Sauter, Zum 'Kleinschen Paradoxon'. Z. Phys. **73**, 547 (1931)
26. W. Heisenberg, H. Euler, Consequences of Dirac theory of the positron. Z. Phys. **98**, 714 (1936)
27. J. Schwinger, On gauge invariance and vacuum polarization. Phys. Rev. **82**, 664 (1951)
28. T. Tajima, G. Mourou, Zettawatt-exawatt lasers and their applications in ultrastrong-field physics. Phys. Rev. ST Accel. Beams **5**, 031301 (2002)
29. A. Di Piazza, C. Müller, K.Z. Hatsagortsyan, C.H. Keitel, Extremely high-intensity laser interactions with fundamental quantum systems. Rev. Mod. Phys **84**, 1177 (2012)
30. F.V. Hartemann, A.L. Troha, H.A. Baldis, A. Gupta, A.K. Kerman, E.C. Landahl, N.C. Luhmann, J.R. Van Meter, High-intensity scattering processes of relativistic electrons in vacuum and their relevance to high-energy astrophysics. Astrophys. J. Suppl. S. **127**, 347 (2000)
31. B.A. Remington, R. Drake, D.D. Ryutov, Experimental astrophysics with high power lasers and Z pinches. Rev. Mod. Phys. **78**, 755 (2006)
32. S.H. Glenzer, R. Redmer, X-ray Thomson scattering in high energy density plasmas. Rev. Mod. Phys. **81**, 1625 (2009)
33. F. Mandl, T.H.R. Skyrme, The theory of the double Compton effect. P. Roy. Soc. Lond. A Mat. **215**, 497 (1952)
34. P.E. Cavanagh, The double Compton effect. Phys. Rev. **87**, 1131 (1952)
35. E. Lötstedt, U.D. Jentschura, Nonperturbative treatment of double compton backscattering in intense laser fields. Phys. Rev. Lett. **103**, 110404 (2009)
36. E. Lötstedt, U.D. Jentschura, Correlated two-photon emission by transitions of Dirac-Volkov states in intense laser fields: QED predictions. Phys. Rev. A **80**, 053419 (2009)
37. P.C.W. Davies, Scalar production in Schwarzschild and Rindler metrics. J. Phys. A - Math. Gen. **8**, 609 (1975)
38. W.G. Unruh, Notes on black-hole evaporation. Phys. Rev. D **14**, 870 (1976)
39. L.C.B. Crispino, A. Higuchi, G.E.A. Matsas, The Unruh effect and its applications. Rev. Mod. Phys. **80**, 787 (2008)
40. P. Chen, T. Tajima, Testing Unruh radiation with ultra intense lasers. Phys. Rev. Lett. **83**, 256 (1999)
41. R. Schützhold, G. Schaller, D. Habs, Tabletop creation of entangled multi-keV photon pairs and the Unruh effect. Phys. Rev. Lett. **100**, 091301 (2008)
42. P.G. Thirolf, D. Habs, A. Henig, D. Jung, D. Kiefer, C. Lang, J. Schreiber, C. Maia, G. Schaller, R. Schützhold, T. Tajima, Signatures of the Unruh effect via high-power, short-pulse lasers. Eur. Phys. J. D **55**, 379 (2009)
43. W. Greiner, B. Müller, J. Rafelski, *Quantum Electrodynamics of Strong Fields* (Springer, Heidelberg, 1985)
44. G. Baur, K. Hencken, D. Trautmann, Electronpositron pair production in ultrarelativistic heavy ion collisions. Phys. Rep. **453**, 1 (2007)

45. T.H. Maiman, Stimulated optical radiation in ruby. Nature **187**, 493 (1960)
46. H.R. Reiss, Absorption of light by light. J. Math. Phys. **3**, 59 (1962)
47. J.J. Klein, B.P. Nigam, Birefringence of the vacuum. Phys. Rev. **135**, B1279 (1964)
48. Z. Bialynicka-Birula, I. Bialynicki-Birula, Nonlinear effects in quantum electrodynamics. Photon propagation and photon splitting in an external field. Phys. Rev. D **2**, 2341 (1970)
49. E. Brezin, C. Itzykson, Polarization phenomena in vacuum nonlinear electrodynamics. Phys. Rev. D **3**, 618 (1971)
50. L.S. Brown, T.W.B. Kibble, Interaction of intense laser beams with electrons. Phys. Rev. **133**, A705 (1964)
51. A.I. Nikishov, V.I. Ritus, Quantum processes in the field of a plane electromagnetic wave and in a constant field.1. Sov. Phys. JETP **19**, 529 (1964)
52. Z. Fried, J.H. Eberly, Scattering of a high-intensity, low-frequency electromagnetic wave by an unbound electron. Phys. Rev. **136**, B871 (1964)
53. I.I. Goldman, Intensity effects in Compton scattering. Phys. Lett. **8**, 103 (1964)
54. T.W.B. Kibble, Frequency shift in high-intensity Compton scattering. Phys. Rev. **138**, B740 (1965)
55. Z. Fried, A. Baker, D. Korff, Comments on intensity-dependent frequency shift in Compton scattering and its possible detection. Phys. Rev. **151**, 1040 (1966)
56. T.W.B. Kibble, Mutual refraction of electrons and photons. Phys. Rev. **150**, 1060 (1966)
57. F. Ehlotzky, Theory of renormalization and Compton scattering in quantum electrodynamics of coherent light of high intensity. Z. Phys. A - Hadron. Nucl. **203**, 119 (1967)
58. Y.B. Zel'dovich, Interaction of free electrons with electromagnetic radiation. Sov. Phys. - Uspekhi **18**, 79 (1975)
59. V.N. Baier, V.M. Katkov, A.I. Milshtein, V.M. Strakhovenko, Contribution to theory of quantum processes in field, of an intense electromagnetic wave. J. Exp. Theo. Phys. **69**, 783 (1975)
60. V.N. Baier, A.I. Milshtein, V.M. Strakhovenko, Interaction between a photon and an intense electromagnetic wave. Sov. Phys. JETP **42**, 961 (1976)
61. V.I. Ritus, Quantum effects of the interaction of elementary particles with an intense electromagnetic field. J. Russ. Laser Res. **6**, 497 (1985)
62. A.I. Nikishov, V.I. Ritus, Pair production by a photon and photon emission by an electron in field of an intense electromagnetic wave and in a constant field. Sov. Phys. JETP **25**, 1135 (1967)
63. H.R. Reiss, Pair production in vacuum by interaction of light with light. Bull. Am. Phys. Soc. **16**, 119 (1971)
64. L.V. Keldysh, Ionization in the field of a strong electromagnetic wave. J. Exp. Theo. Phys. **47**, 1945 (1964)
65. H.R. Reiss, Effect of an intense electromagnetic field on a weakly bound system. Phys. Rev. A **22**, 1786 (1980)
66. P.B. Corkum, N.H. Burnett, F. Brunel, Above-threshold ionization in the long-wavelength limit. Phys. Rev. Lett. **62**, 1259 (1989)
67. J.H. Eberly, H.R. Reiss, Electron self-energy in intense plane-wave field. Phys. Rev. **145**, 1035 (1966)
68. H.R. Reiss, J.H. Eberly, Green's function in intense-field electrodynamics. Phys. Rev. **151**, 1058 (1966)
69. Vachaspati, Harmonics in the scattering of light by free electrons. Phys. Rev. **128**, 664 (1962)
70. J.H. Eberly, A. Sleeper, Trajectory and mass shift of a classical electron in a radiation pulse. Phys. Rev. **176**, 1570 (1968)
71. E.S. Sarachik, G.T. Schappert, Classical theory of the scattering of intense laser radiation by free electrons. Phys. Rev. D **1**, 2738 (1970)
72. J.W. Meyer, Covariant classical motion of electron in a laser beam. Phys. Rev. D **3**, 621 (1971)
73. Y.I. Salamin, F.H.M. Faisal, Harmonic generation by superintense light scattering from relativistic electrons. Phys. Rev. A **54**, 4383 (1996)
74. Y.I. Salamin, F.H.M. Faisal, Ponderomotive scattering of electrons in intense laser fields. Phys. Rev. A **55**, 3678 (1997)

75. F.V. Hartemann, D.J. Gibson, A.K. Kerman, Classical theory of Compton scattering: assessing the validity of the Dirac-Lorentz equation. Phys. Rev. E **72**, 026502 (2005)
76. D. Strickland, G. Mourou, Compression of amplified chirped optical pulses. Opt. Comm. **56**, 219 (1985)
77. P. Maine, D. Strickland, P. Bado, M. Pessot, G. Mourou, Generation of ultrahigh peak power pulses by chirped pulse amplification. IEEE J. Quant. Elect. **24**, 398 (1988)
78. M.D. Perry, G. Mourou, Terawatt to petawatt subpicosecond lasers. Science **264**, 917 (1994)
79. V. Yanovsky, V. Chvykov, G. Kalinchenko, P. Rousseau, T. Planchon, T. Matsuoka, A. Maksimchuk, J. Nees, G. Cheriaux, G. Mourou, K. Krushelnick, Ultra-high intensity-300-TW laser at 0.1 Hz repetition rate. Opt. Exp. **16**, 2109 (2008)
80. PFS, Petawatt Field Synthesizer, http://www.attoworld.de/Home/ourResearch/ToolsAndTechniques/Petawatt-scaleSourceOfIrLight/index.html
81. ELI, Extreme Light Infrastructure, http://www.extreme-light-infrastructure.eu/
82. HiPER, High Power laser Energy Research facility, http://www.hiper-laser.org/index.asp/
83. G. Mourou, T. Tajima, More intense shorter pulses. Science **331**, 41 (2011)
84. G. Mourou, T. Tajima, S.V. Bulanov, Optics in the relativistic regime. Rev. Mod. Phys. **78**, 309 (2006)
85. L.D. Landau, E.M. Lifschitz, *Lehrbuch der Theoretischen Physik - Band 2: Klassische Feldtheorie* (Akademie Verlag, Berlin, 1997)
86. H.R. Reiss, Proposed experiment to detect the mass shift of an electron in an intense photon field. Phys. Rev. Lett. **17**, 1162 (1966)
87. C. Harvey, T. Heinzl, A. Ilderton, M. Marklund, The intensity dependent mass shift: existence, universality and detection. Phys. Rev. Lett. **109**, 100402 (2012)
88. J.H. Eberly, Proposed experiment for observation of nonlinear compton wavelength shift. Phys. Rev. Lett. **15**, 91 (1965)
89. C. Bamber, S.J. Boege, T. Koffas, T. Kotseroglou, A.C. Melissinos, D.D. Meyerhofer, D.A. Reis, W. Ragg, C. Bula, K.T. McDonald, E.J. Prebys, D.L. Burke, R.C. Field, G. Horton-Smith, J.E. Spencer, D. Walz, S.C. Berridge, W.M. Bugg, K. Shmakov, A.W. Weidemann, Studies of nonlinear QED in collisions of 46.6 GeV electrons with intense laser pulses. Phys. Rev. D **60**, 092004 (1999)
90. T.J. Englert, E.A. Rinehart, Second-harmonic photons from the interaction of free electrons with intense laser radiation. Phys. Rev. A **28**, 1539 (1983)
91. C. Bula, K.T. McDonald, E.J. Prebys, C. Bamber, S. Boege, T. Kotseroglou, A.C. Melissinos, D.D. Meyerhofer, W. Ragg, D.L. Burke, R.C. Field, G. Horton-Smith, A.C. Odian, J.E. Spencer, D. Walz, S.C. Berridge, W.M. Bugg, K. Shmakov, A.W. Weidemann, Observation of nonlinear effects in Compton scattering. Phys. Rev. Lett. **76**, 3116 (1996)
92. S.Y. Chen, A. Maksimchuk, D. Umstadter, Experimental observation of relativistic nonlinear Thomson scattering. Nature **396**, 653 (1998)
93. D.L. Burke, R.C. Field, G. Horton-Smith, J.E. Spencer, D. Walz, S.C. Berridge, W.M. Bugg, K. Shmakov, A.W. Weidemann, C. Bula, K.T. McDonald, E.J. Prebys, C. Bamber, S.J. Boege, T. Koffas, T. Kotseroglou, A.C. Melissinos, D.D. Meyerhofer, D.A. Reis, W. Ragg, Positron production in multiphoton light-by-light scattering. Phys. Rev. Lett. **79**, 1626 (1997)
94. V.N. Baier, V.M. Katkov, V.M. Strakhovenko, *Electromagnetic Processes at High Energies in Oriented Single Crystals* (World Scientific, Singapore, 1994)
95. A.H. Compton, A quantum theory of the scattering of X-rays by light elements. Phys. Rev. **21**, 483 (1923)
96. T. Tajima, J.M. Dawson, Laser electron accelerator. Phys. Rev. Lett. **43**, 267 (1979)
97. C.G.R. Geddes, C. Toth, J. van Tilborg, E. Esarey, C.B. Schroeder, D. Bruhwiler, C. Nieter, J. Cary, W.P. Leemans, High-quality electron beams from a laser wakefield accelerator using plasma-channel guiding. Nature **431**, 538 (2004)
98. W.P. Leemans, B. Nagler, A.J. Gonsalves, C. Toth, K. Nakamura, C.G.R. Geddes, E. Esarey, C.B. Schroeder, S.M. Hooker, GeV electron beams from a centimetre-scale accelerator. Nat. Phys. **2**, 696 (2006)

99. C.E. Clayton, J.E. Ralph, F. Albert, R.A. Fonseca, S.H. Glenzer, C. Joshi, W. Lu, K.A. Marsh, S.F. Martins, W.B. Mori, A. Pak, F.S. Tsung, B.B. Pollock, J.S. Ross, L.O. Silva, D.H. Froula, Self-guided laser wakefield acceleration beyond 1 GeV using ionization-induced injection. Phys. Rev. Lett. **105**, 105003 (2010)
100. Y.I. Salamin, Z. Harman, C.H. Keitel, Direct high-power laser acceleration of ions for medical applications. Phys. Rev. Lett. **100**, 155004 (2008)
101. R. Kodama, P.A. Norreys, K. Mima, A.E. Dangor, R.G. Evans, H. Fujita, Y. Kitagawa, K. Krushelnick, T. Miyakoshi, N. Miyanaga, T. Norimatsu, S.J. Rose, T. Shozaki, K. Shigemori, A. Sunahara, M. Tampo, K.A. Tanaka, Y. Toyama, Y. Yamanaka, M. Zepf, Fast heating of ultrahigh-density plasma as a step towards laser fusion ignition. Nature **412**, 798 (2001)
102. R. Kodama, H. Shiraga, K. Shigemori, Y. Toyama, S. Fujioka, H. Azechi, H. Fujita, H. Habara, T. Hall, Y. Izawa, T. Jitsuno, Y. Kitagawa, K. Krushelnick, K.L. Lancaster, K. Mima, K. Nagai, M. Nakai, H. Nishimura, T. Norimatsu, P.A. Norreys, S. Sakabe, K.A. Tanaka, A. Youssef, M. Zepf, Fast ignitor consortium, Nuclear fusion—Fast heating scalable to laser fusion ignition. Nature **418**, 933 (2002)
103. E. Esarey, S.K. Ride, P. Sprangle, Nonlinear Thomson scattering of intense laser pulses from beams and plasmas. Phys. Rev. E **48**, 3003 (1993)
104. Y.Y. Lau, F. He, D. Umstadter, R. Kowalczyk, Nonlinear Thomson scattering: a tutorial. Phys. Plas. **10**, 2155 (2003)
105. B. Liesfeld, *A Photon Collider at Relativistic Intensity* (Books on Demand, Norderstedt, 2006)
106. C. Müller, K.Z. Hatsagortsyan, C.H. Keitel, Particle physics with a laser-driven positronium atom. Phys. Lett. B **659**, 209 (2008)
107. C. Müller, A. Di Piazza, A. Shahbaz, T. Bürvenich, J. Evers, K.Z. Hatsagortsyan, C.H. Keitel, High-energy, nuclear, and QED processes in strong laser fields. Laser Phys. **18**, 175 (2008)
108. C. Müller, C. Deneke, C.H. Keitel, Muon-pair creation by two X-ray laser photons in the field of an atomic nucleus. Phys. Rev. Lett. **101**, 060402 (2008)
109. A. Di Piazza, K.Z. Hatsagortsyan, C.H. Keitel, Quantum radiation reaction effects in multiphoton Compton scattering. Phys. Rev. Lett. **105**, 220403 (2010)
110. A. Di Piazza, K.Z. Hatsagortsyan, C.H. Keitel, Harmonic generation from laser-driven vacuum. Phys. Rev. D **72**, 085005 (2005)
111. A. Di Piazza, K.Z. Hatsagortsyan, C.H. Keitel, Light diffraction by a strong standing electromagnetic wave. Phys. Rev. Lett. **97**, 083603 (2006)
112. A. Di Piazza, K.Z. Hatsagortsyan, C.H. Keitel, Nonlinear interaction of strong laser fields in vacuum. Laser Phys. **17**, 345 (2007)
113. B. King, A. Di Piazza, C.H. Keitel, A matterless double slit. Nature Photonics **4**, 92 (2010)
114. H.G. Hetzheim, C.H. Keitel, Ionization dynamics versus laser intensity in laser-driven multiply charged ions. Phys. Rev. Lett. **102**, 083003 (2009)
115. O. Har-Shemesh, A. Di Piazza, Peak intensity measurement of relativistic lasers via nonlinear Thomson scattering. Opt. Exp. **37**, 1352 (2012)
116. F. Mackenroth, A. Di Piazza, C.H. Keitel, Determining the carrier-envelope phase of intense few-cycle laser pulses. Phys. Rev. Lett. **105**, 063903 (2010)
117. A. Zeilinger, Experiment and the foundations of quantum physics. Rev. Mod. Phys. **71**, S288 (1999)
118. A. Peres, D.R. Terno, Quantum information and relativity theory. Rev. Mod. Phys. **76**, 93 (2004)
119. H.C. Wu, J. Meyer-ter-Vehn, Giant half-cycle attosecond pulses. Nat. Photonics **6**, 304 (2012)
120. A. Bonvalet, M. Joffre, J.L. Martin, A. Migus, Generation of ultrabroadband femtosecond pulses in the mid-infrared by optical rectification of 15 fs light pulses at 100 MHz repetition rate. Appl. Phys. Lett. **67**, 2907 (1995)
121. G. Krauss, S. Lohss, T. Hanke, A. Sell, S. Eggert, R. Huber, A. Leitenstorfer, Synthesis of a single cycle of light with compact erbium-doped fibre technology. Nat. Photonics **4**, 33 (2010)
122. A.L. Cavalieri, E. Goulielmakis, B. Horvath, W. Helml, M. Schultze, M. Fie, V. Pervak, L. Veisz, V.S. Yakovlev, M. Uiberacker, A. Apolonski, F. Krausz, R. Kienberger, Intense 1.5-cycle near infrared laser waveforms and their use for the generation of ultra-broadband soft-x-ray harmonic continua. New. J. Phys. **9**, 242 (2007)

123. G. Sansone, E. Benedetti, F. Calegari, C. Vozzi, L. Avaldi, R. Flammini, L. Poletto, P. Villoresi, C. Altucci, R. Velotta, S. Stagira, S. De Silvestri, M. Nisoli, Isolated single-cycle attosecond pulses. Science **314**, 443 (2006)
124. E. Goulielmakis, M. Schultze, M. Hofstetter, V.S. Yakovlev, J. Gagnon, M. Uiberacker, A.L. Aquila, E.M. Gullikson, D.T. Attwood, R. Kienberger, F. Krausz, U. Kleineberg, Single-cycle nonlinear optics. Science **320**, 1614 (2008)
125. N.B. Narozhnyi, M.S. Fofanov, Photon emission by electron during collision with short focused laser pulses. J. Exp. Theo. Phys. **110**, 26 (1996)
126. P.J. Mohr, B.N. Taylor, D.B. Newell, CODATA recommended values of the fundamental physical constants: 2006. Rev. Mod. Phys. **80**, 633 (2008)

Chapter 2
Interaction of Electrons with Laser Fields

To describe the emission by electrons, it is necessary to first analyze their dynamics. There are several theories available, as to how this task is to be accomplished. In this chapter we are going to outline two fundamentally different approaches. We will start by presenting the classical framework, in which an electron of charge e is viewed as a point source of an electromagnetic field. The emission of such a classical current in discussed in Sect. 2.1.

As discussed in Chap. 1, in the past century it emerged a quantum theory of matter. In this theory, due to the intrinsic uncertainty of conjugate observables, in particular position and momentum, no point particles can exist. Much rather, quantum electrodynamics describes particles as exited states of quantized fields and describes a scattering as the probability of a quantum state, formed in the far past, to go over into another quantum state, observed in the far future. A short introduction into the schemes and techniques of the according scattering theory of quantum electrodynamics is given in Sect. 2.2.

2.1 Classical Electrodynamics

The basics of classical electrodynamics are the famous Maxwell equations, unifying the electric and magnetic interaction. In covariant form they state that a spatial charge current $\boldsymbol{j}$, which is combined with its charge density ρ to a four dimensional current $j^\mu = (c\rho, \boldsymbol{j})$, will generate electromagnetic fields according to [1]

$$\partial_\mu F^{\mu\nu}(x) = \frac{4\pi}{c} j^\nu(x), \tag{2.1}$$

where $F^{\mu\nu}(x) = \partial^\mu A^\nu(x) - \partial^\nu A^\mu(x)$ is the antisymmetric field strength tensor derived from the four potential $A^\mu(x) = (\phi(x), \boldsymbol{A}(x))$. This potential, however, is not a measurable quantity as it is not uniquely defined by Eq. (2.1). In fact, applying a gauge transformation to the four potential, the field strength tensor, entering the

K. F. Mackenroth, *Quantum Radiation in Ultra-Intense Laser Pulses*,
Springer Theses, DOI: 10.1007/978-3-319-07740-6_2,

Maxwell equations, remains unchanged. The same current $j^\mu(x)$ thus generates a whole equivalence class of vector potentials, all connected by gauge freedom. The electromagnetic quantities, uniquely defined by Eq. (2.1) are the three dimensional vector fields of the electric and magnetic field $\boldsymbol{\mathcal{E}}(x)$ and $\boldsymbol{\mathcal{B}}(x)$, respectively, which are derived from the four potential via the relations

$$\boldsymbol{\mathcal{E}}(x) = -\nabla\phi(x) - \frac{\partial \boldsymbol{A}(x)}{\partial t} \tag{2.2a}$$

$$\boldsymbol{\mathcal{B}}(x) = \nabla \times \boldsymbol{A}(x). \tag{2.2b}$$

For describing electromagnetic fields propagating in free space ($j^\nu \equiv 0$) one usually adopts the Lorenz gauge $\partial_\mu A^\mu(x) = \partial_t\phi(x) + \nabla\boldsymbol{A}(x) = 0$, whence Eq. (2.1) turns into a wave equation of the form

$$\Box A^\mu(x) = 0, \tag{2.3}$$

where the D'Alembert operator $\Box = \partial_\mu\partial^\mu = \partial_t^2 - \nabla^2$ is introduced. As we wish to describe physical fields propagating in free space, the employed solutions have to satisfy Eq. (2.3). An important class of such solutions is given by plane waves, which are defined by a wave vector $k_L^\mu = \omega_L(1, \boldsymbol{n}_L)$, with the central angular frequency of the electric field ω_L, and depend on the spatial coordinates only via the so-called invariant phase $\eta = x_\mu k_L^\mu$ [1]. Much of the discussion presented in this work will be based on linearly polarized plane wave solutions of Eq. (2.3), for which we will use the notation $A_L^\mu(\eta) = A_L\epsilon_L^\mu\psi_A(\eta)$, with the constant (positive) amplitude $A_L = -m\xi/e$, the wave's polarization four vector ϵ_L^μ and the shape function $\psi_A(\eta)$ encoding the temporal structure of the field. It is customary to consider plane waves, propagating in free space, in a reference system, where the static potential vanishes ($\phi \equiv 0$) and all physical fields are derived from the vector potential $\boldsymbol{A}_L(\eta)$. This corresponds to a purely spatial polarization vector $\epsilon_L^\mu = (0, \boldsymbol{\epsilon}_L)$. For a plane wave field the Lorenz gauge condition reduces to $A_L k_L = 0$, whence with the above choices we find

$$\boldsymbol{\epsilon}_L\boldsymbol{k}_L = 0. \tag{2.4}$$

A plane wave thus is always polarized perpendicularly to its propagation direction. The shape function is an essential ingredient in obtaining specific results. It is customary to model it by a function of the form $\psi_A(\eta) = g(\eta)\sin(\eta + \eta_0)$, where $g(\eta)$ is the so-called *pulse envelope*, which has a peak value of unity. The validity of this approach has been proven on the basis that the central carrying frequency is uniquely defined by the envelope. In particular, it needs to be independent of the quantity η_0, which quantifies a relative phase shift between the carrier wave and the envelope and is hence labeled *carrier-envelope phase (CEP)* [2]. Due to the importance of the shape function, we wish to explicitly present two possible choices here. Here we are going to present the shape functions as functions of the invariant laser phase η, but also point out that by virtue of the relation $\eta = k_L^+ x^-$ any function $F(\eta)$ is easily translated to a function of the light cone coordinate x^- according to

$$F(x^-) := F(k_L^+ x^-). \tag{2.5}$$

This equivalence is frequently used for the shape function and the corresponding expression for the four potential $A_L^\mu(x^-) = A_L \epsilon_L^\mu \psi_{\mathcal{A}}(x^-)$ in the course of this thesis, but for the sake of notational simplicity we are not going to introduce a separate symbol for these two, rigorously speaking different, functions of Eq. (2.5). To model a few-cycle pulse we use

$$\psi_{\mathcal{A}}(\eta) = \begin{cases} \sin^4\left(\frac{\eta}{2n_C}\right)\sin(\eta+\eta_0) & \text{if } \eta \in [0, 2\pi\, n_C] \\ 0 & \text{else,} \end{cases} \tag{2.6}$$

where n_C is the number of cycles contained in the laser pulse. The favors of this choice are its simple analytic structure alongside its smooth rise and fall of the electric field, derived from Eq. (2.2a), which is well suited to model a laser pulse [3]. One drawback is that the average of the envelope function (2.6) is constantly smaller than one, irrespective of the number of cycles contained in the laser pulse. This can be seen by the computation

$$\langle g \rangle = \frac{\int_0^{2\pi\, n_C} d\eta\, \sin^4\left(\frac{\eta}{2n_C}\right)}{2\pi\, n_C} = \frac{\int_0^{2\pi} d\eta'\, \sin^4\left(\frac{\eta'}{2}\right)}{2\pi} = \frac{3}{8}. \tag{2.7}$$

Despite the good applicability of Eq. (2.6) to model few-cycle pulses, due to the sketched drawback, it is problematic, in case one wants to model a long laser pulse, or particularly recover the monochromatic limit. This limit namely is recovered, if the laser pulse can approximately be described by the sole oscillation frequency. The amplitude variation over the whole pulse must accordingly be negligible, which in turn, for a shape function normalized to unity, implies that Eq. (2.7) has to equal unity. To recover the monochromatic limit or model a longer laser pulse we thus employ the following, however more complicated envelope function

$$g^{\text{long}}(\eta) = \begin{cases} \frac{\eta}{2\pi\, n_{\text{switch}}} & \text{if } \eta \in [0, 2\pi\, n_{\text{switch}}] \\ 1 & \text{if } \eta \in [2\pi\, n_{\text{switch}}, 2\pi\,(n_{\text{switch}} + n_{\text{flat}})] \\ \frac{(2n_{\text{switch}} + n_{\text{flat}}) - \eta/2\pi}{n_{\text{switch}}} & \text{if } \eta \in [2\pi\,(n_{\text{switch}} + n_{\text{flat}}), 2\pi\,(2n_{\text{switch}} + n_{\text{flat}})] \\ 0 & \text{else.} \end{cases} \tag{2.8}$$

To obtain a proper shape function $\psi_{\mathcal{A}}(\eta)$, one has to multiply this envelope function with the oscillating carrier function $\sin(\eta+\eta_0)$, analogous to Eq. (2.6). That Eq. (2.8) indeed allows to take the monochromatic limit, is seen in analogy to Eq. (2.7)

$$\langle g^{\text{long}} \rangle = \frac{\int_0^{2\pi n_C} d\eta\, g^{\text{long}}(\eta)}{2\pi\,(2n_{\text{switch}} + n_{\text{flat}})} = \frac{n_{\text{flat}} + n_{\text{switch}}}{2n_{\text{switch}} + n_{\text{flat}}} \xrightarrow{n_{\text{flat}}\to\infty} 1. \tag{2.9}$$

In the light of the previous computations please note that, due to the relation $\boldsymbol{\mathcal{E}}_L(\eta) = -\partial_t \boldsymbol{A}_L(\eta)$, the value of the four potential can be viewed as

$$\boldsymbol{A}_L(\eta) = -\int_{-\infty}^{\eta} d\eta' \, \frac{dt}{d\eta'} \boldsymbol{\mathcal{E}}_L(\eta'), \tag{2.10}$$

where any integration constant can be chosen as zero by gauge freedom. The above expression, however, for $\eta \to \infty$ is proportional to the zero-frequency, i.e. constant field, Fourier component of the laser's electric field. Since such a constant field mode, however, does not propagate, it is essential that for any choice of $\psi_{\mathcal{A}}(\eta)$ it holds $A_L(\eta \to \infty) = 0$. Consequently, even though $|A_L(\eta \to \infty)| > 0$ would correspond to a physically reasonable electric field vanishing at infinity, this possibility is ruled out.

From here on the following discussion is again valid for arbitrary electromagnetic fields fulfilling Eq. (2.3). To determine the dynamics of an electron moving inside such an electromagnetic field, one has to solve its equation of motion. Any external electromagnetic field described by its field strength tensor $F_L^{\mu\nu}(x)$, obtained via Eq. (2.3), exerts a force on the electron according to the Lorentz force equation [1]

$$\frac{dp^\mu(s)}{ds} = \frac{e}{m} F_L^{\mu\nu}(x) p_\nu(s). \tag{2.11}$$

In this expression s is the proper time of the electron and $p^\mu(s)$ its kinetic momentum. Integrating Eq. (2.11) together with the electron's initial momentum $\boldsymbol{p}_i$ and position $\boldsymbol{x}_i$ then yields the classical trajectory of the electron subjected to the electromagnetic field in question. To be able to show a specific example, however, we first have to specify the coordinate frame, in which we wish to investigate the interaction. We introduce the reference frame, in which we will observe the interaction of a laser pulse with an electron throughout this thesis in Fig. 2.1. We are going to consider the laser's propagation and polarization axes to be the z- and x-axis, respectively. This corresponds to the representations $\boldsymbol{k}_L = (0, 0, 1)$ and $\epsilon_L = (1, 0, 0)$. The coordinate frame will be chosen such that the electron is initially counterpropagating to the laser pulse ($\boldsymbol{p}_i = \varepsilon_i(0, 0, -\beta_i)$), where β_i is the electron's initial velocity. The observation direction of the electron radiation, observed in this reference frame will be denoted by a direction vector $\boldsymbol{n}_1 = (\sin(\vartheta_1)\cos(\varphi_1), \sin(\vartheta_1)\sin(\varphi_1), \cos(\vartheta_1))$, where ϑ_1 is the angle between $\boldsymbol{n}_1$ and $\boldsymbol{n}_L$ and φ_1 is the angle between $\boldsymbol{n}_1$ and the x-z plane. Exemplary electron trajectories, obtained via numerical integration of Eq. (2.11) and observed in a reference frame according to Fig. 2.1, are shown in Fig. 2.2. The trajectory inside a circularly polarized laser wave in Fig. 2.2b is shown for mere comparison,

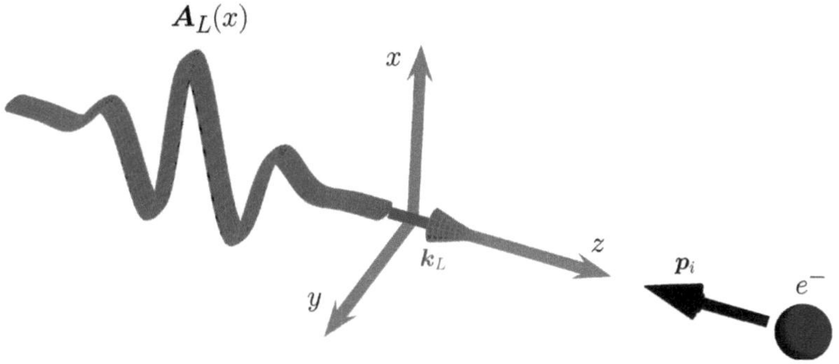

Fig. 2.1 Generic choice for the reference frame the scattering will be observed in

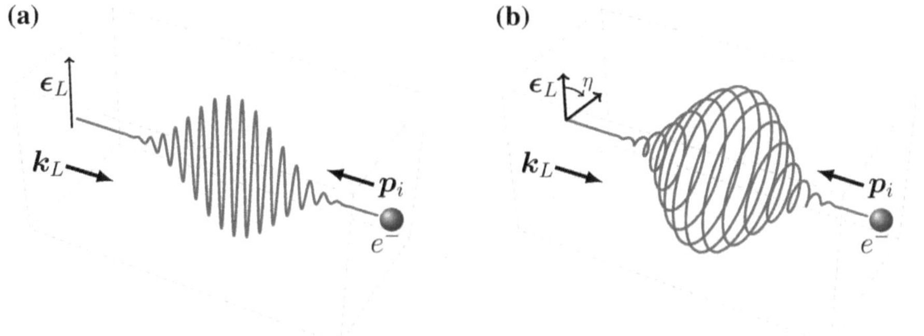

Fig. 2.2 Classical trajectories of an electron colliding head on with a laser pulse. **a** Exemplary electron trajectory for propagation in a linearly polarized laser beam. **b** Exemplary electron trajectory for propagation in a circularly polarized laser beam

as we are going to consider exclusively linearly polarized laser pulses in this work. As seen in Fig. 2.2a, in this case the classical trajectory is confined to the $\boldsymbol{k}_L$-ϵ_L-plane. It is then sufficient to give the two dimensional electron trajectory within this plane to fully describe the electron's classical dynamics. From here on we will adopt this simplified visualization scheme in the course of this work. Please note that by employing Eq. (2.11) in this work, we are going to neglect any influence of the electron's self-field on its own dynamics. Such radiation reaction, however, was shown to possibly significantly influence the dynamics [4], and is a matter of intense scientific discourse [5–7].

Solving Eq. (2.11) results in a given trajectory for the electron $r^\mu(t) = (t, \boldsymbol{r}(t))$ with a velocity $u^\mu(t) = \partial_s r^\mu(t) = \gamma(t)\beta^\mu(t)$, where $\gamma(t) = \varepsilon(t)/m$ is the electron's relativistic factor, s its proper time and $\beta^\mu(t) = (1, \boldsymbol{\beta}(t))$. From this quantity one defines the classical charge current of an electron $j^\mu(x) = e\beta^\mu(t)\delta(\boldsymbol{x} - \boldsymbol{r}(t))$. The electron's emission can then be computed by means of the Lienard Wiechert potentials [1, 8]. According to this formalism the energy emitted by an accelerated point-like electron into the observation direction $\boldsymbol{n}$ per unit frequency and solid angle element is given by

$$\frac{dE}{d\omega d\Omega} = \frac{e^2\omega^2}{4\pi^2} \left| \int_{-\infty}^{\infty} dt\, \boldsymbol{n} \times (\boldsymbol{n} \times \boldsymbol{\beta}(t))\, \mathrm{e}^{\mathrm{i}\omega\,(t-\boldsymbol{nr}(t))} \right|^2 . \tag{2.12}$$

The integrand is simplified through

$$\boldsymbol{n} \times (\boldsymbol{n} \times \boldsymbol{\beta}(t)) = \boldsymbol{n}\,(\boldsymbol{n}\boldsymbol{\beta}(t)) - \boldsymbol{\beta}(t). \tag{2.13}$$

To simplify the term containing the factor $\boldsymbol{n}\boldsymbol{\beta}(t)$ one writes

$$\int_{-\infty}^{\infty} dt\ (1-\boldsymbol{n}\boldsymbol{\beta}(t))\ \mathrm{e}^{\mathrm{i}\omega\,(t-\boldsymbol{nr}(t))} = \int_{-\infty}^{\infty} dt \left(\frac{d}{dt} \frac{\mathrm{e}^{\mathrm{i}\omega \int_{-\infty}^{t} dt'\ (1-\boldsymbol{n}\boldsymbol{\beta}(t'))}}{\mathrm{i}\omega} \right) = 0. \tag{2.14}$$

The transformed integral has to vanish since its integrand is the total differential of an expression, which does not contribute at $t = \pm\infty$. Hence, for Eq. (2.12) one obtains

$$\frac{dE}{d\omega d\Omega} = \frac{e^2\omega^2}{4\pi^2} \left| \int_{-\infty}^{\infty} dt\ (\boldsymbol{n} - \boldsymbol{\beta}(t))\mathrm{e}^{\mathrm{i}\omega\,(t-\boldsymbol{nr}(t))} \right|^2 . \tag{2.15}$$

Taking the square in this expression results in

$$\begin{aligned} \frac{dE}{d\omega d\Omega} &= \frac{e^2\omega^2}{4\pi^2} \int_{-\infty}^{\infty} dtdt'\ (1 - \boldsymbol{n}\boldsymbol{\beta}(t) - \boldsymbol{n}\boldsymbol{\beta}(t') + \boldsymbol{\beta}(t)\boldsymbol{\beta}(t')) \\ &\quad \times \mathrm{e}^{\mathrm{i}\omega\,(t-\boldsymbol{nr}(t))}\mathrm{e}^{\mathrm{i}\omega\,(t'-\boldsymbol{nr}(t'))} \\ &= \frac{e^2\omega^2}{4\pi^2} \int_{-\infty}^{\infty} dtdt'\ (\boldsymbol{\beta}(t)\boldsymbol{\beta}(t') - 1)\mathrm{e}^{\mathrm{i}\omega\,(t-\boldsymbol{nr}(t))}\mathrm{e}^{\mathrm{i}\omega\,(t'-\boldsymbol{nr}(t'))}, \end{aligned} \tag{2.16}$$

where, for obtaining the second line, Eq. (2.14) is used. Now, Eq. (2.16) is easily written in a covariant form, recalling the four velocity $u^\mu(t) = p^\mu(t)/m$. Plugging this into Eq. (2.16) we find the covariant expression [9]

$$\frac{dE}{d\omega d\Omega} = \frac{e^2\omega^2}{4\pi^2} \left| \int_{-\infty}^{\infty} dt\ \frac{p^\mu(t)}{\varepsilon(t)} \mathrm{e}^{ikr(t)} \right|^2 , \tag{2.17}$$

where we defined $k^\mu = \omega(1, \boldsymbol{n})$ in the exponential. Please note that the appearance of $\varepsilon(t)$ does not make this expression change under Lorentz transformation since it is integrated over the time t, which compensates the transformation of the energy.

We finally wish to discuss some qualitative features of the angular and frequency distribution that is expected in the radiation from an electron scattered from a laser field (see [1, 8]). The former distribution is dominated by the properties of the Lorentz transformations, as we wish to briefly sketch. Consider an electron which propagates with the instantaneous velocity four vector $\beta^\mu = (1, 0, 0, \beta^z)$, observed in a laboratory frame according to Fig. 2.1. In a reference frame, copropagating with the electron at β^z, the radiation emitted by the electron will feature a wave vector $k'^\mu = \omega'(1, \sin(\vartheta'), 0, \cos(\vartheta'))$, where ϑ' is the angle between $\boldsymbol{k}'$ and the negative z-axis measured in the copropagating reference frame. Additionally the perpendicular axes are chosen such that the emission is confined to the x-z-plane. Transforming this four vector into the laboratory frame, where the emitted radiation is observed, one obtains

$$k^\mu = \omega \begin{pmatrix} 1 \\ \sin(\vartheta) \\ 0 \\ \cos(\vartheta) \end{pmatrix} = \frac{\omega'}{m} \begin{pmatrix} \varepsilon(1 + \beta^z \cos(\vartheta')) \\ m \sin(\vartheta') \\ 0 \\ \varepsilon(\beta^z + \cos(\vartheta')) \end{pmatrix}, \tag{2.18}$$

where ϵ is the electron's energy, measured in the laboratory frame. The angle between the wave vector and the electron's propagation direction in the laboratory frame thus becomes

$$\sin(\vartheta) = \frac{m}{\varepsilon} \frac{\sin(\vartheta')}{(1 + \beta^z \cos(\vartheta'))} \sim \frac{m}{\varepsilon} \tag{2.19}$$

One concludes that the emission of an electron in highly relativistic motion ($\varepsilon \gg m$) is confined to a narrow cone of opening angle $\Delta\vartheta \sim m/\varepsilon$ around its velocity vector at the time of emission. Any observer detecting the emission from the electron will then detect only a short burst of radiation, whenever the electron's velocity points into his observation direction. Over such short times the change of the electron's propagation direction can be approximated by a circular orbit with the instantaneous radius of curvature ρ. The time in the highly relativistic electron's rest frame, over which its emission cone will accordingly illuminate a detector, then scales as

$$\Delta s \sim \frac{m\rho}{\varepsilon}. \tag{2.20}$$

The transformation of this illumination time to the observation time t, measured in the laboratory in which the radiation is observed, requires the Lorentz transformation factor

$$\frac{dt}{ds} = 1 - \boldsymbol{n\beta} \sim \left(\frac{m}{\varepsilon}\right)^2, \tag{2.21}$$

where $\boldsymbol{n}$ again is a unit vector, pointing along the direction of observation. A laboratory detector will thus detect a radiation flash of the approximate duration

$$\Delta t \sim \rho \left(\frac{m}{\varepsilon}\right)^3, \tag{2.22}$$

whenever the electron points in its direction. According to the general theory of Fourier transformation, a field flash of such short duration has to contain frequencies up to

$$\omega_c \sim \frac{1}{\rho}\left(\frac{\varepsilon}{m}\right)^3. \tag{2.23}$$

One thus expects the radiation of an electron scattered by an intense laser pulse to scale as the cube of its instantaneous energy.

2.1.1 Electron Radiation in a Plane Wave

The classical equation of motion was solved analytically for the momentum in the case of an electron moving in a plane wave laser field exactly by solving the Hamilton-Jacobi equation [10–12] as well as by direct integration [13, 14]. In the former case the phase dependent position of the electron is found directly as derivative of the action (or alternatively principal function) of an electron entering a plane wave field $A_L^\mu(\eta)$ with a momentum $p_i = p(\eta \to -\infty)$ [1]

$$S_{p_i}(x) = -p_i x - \int_0^\eta \mathrm{d}\phi \left(e\frac{p_i A_L(\phi)}{p_i k_L} - \frac{e^2 A_L^2(\phi)}{2(p_i k_L)^2} \right). \tag{2.24}$$

In the latter approach it is obtained by another integration of the equation of motion (2.11) and $u^\mu(t) = dr^\mu(t)/ds$. An advantage of the latter computation is that its results are given in an explicitly covariant manner which is why we sketch this way of solving the classical equations of motion. As a first step we note that for a plane wave electromagnetic field the field strength tensor satisfies $F_L^{\mu\nu}(\eta) = k_L^\mu \partial_\eta A^\nu(\eta) - k_L^\nu \partial_\eta A^\mu(\eta)$. From the classical equation of motion (2.11) one concludes

$$\frac{dp^\mu(s)}{ds} = \frac{e}{m}\left(k_L^\mu (A_L p(s)) - A_L^\mu (p(s)k_L)\right) \partial_\eta \psi_A(\eta). \tag{2.25}$$

Multiplying Eq. (2.25) with the constant plane wave's wave vector we conclude

$$\frac{d(p(\eta)k_L)}{ds} = 0, \tag{2.26}$$

where we utilized the gauge condition $k_L A_L = 0$ and whence we conclude that $p(s)k_L = p_i k_L$ is a constant of motion and can thus be set to its initial value. Furthermore, for the laser's invariant phase one finds

$$\frac{d\eta}{ds} = k_L^\mu \frac{dx_\mu}{ds} = \frac{p_i k_L}{m}. \tag{2.27}$$

It is then advantageous to parameterize the electron's kinetic momentum by η and one can change the variable in Eq. (2.25) according to

$$\begin{aligned}\frac{dp^\mu(\eta)}{d\eta} &= \frac{ds}{d\eta}\frac{dp^\mu(s)}{ds} \\ &= \frac{e}{p_i k_L}\left(k_L^\mu (A_L p(\eta)) - A_L^\mu (p_i k_L)\right)\partial_\eta \psi_{\mathcal{A}}(\eta).\end{aligned} \tag{2.28}$$

Multiplying Eq. (2.28) with the constant plane wave's amplitude vector A_L^μ we find

$$\begin{aligned}&\frac{d(A_L p(\eta))}{d\eta} = -eA_L^\mu \partial_\eta A_L^\mu(\eta) \\ &\Rightarrow (A_L p(\eta)) = (A_L p_i) - eA_L^2 \psi_{\mathcal{A}}(\eta).\end{aligned} \tag{2.29}$$

Inserting now Eq. (2.26) and (2.29) into Eq. (2.28) we find

$$\frac{dp^\mu(\eta)}{d\eta} = e\left(k_L^\mu \frac{(A_L p_i) - eA_L^2 \psi_{\mathcal{A}}(\eta)}{p_i k_L} - A_L^\mu\right)\partial_\eta \psi_{\mathcal{A}}(\eta), \tag{2.30}$$

which is readily integrated to give the covariant form of an electron's momentum in the presence of a plane wave laser field as

$$p^\mu(\eta) = p_i^\mu - eA^\mu(\eta) + k_L^\mu \left[e\frac{p_i A_L(\eta)}{p_i k_L} - \frac{e^2 A_L^2(\eta)}{2(p_i k_L)}\right] \tag{2.31a}$$

$$r^\mu(\eta) = r_i + \int_{-\infty}^{\eta} d\eta' \, \frac{p_i^\mu - eA^\mu(\eta')}{p_i k_L} + k^\mu \int_{-\infty}^{\eta} d\eta' \, \frac{e(A_L(\eta')p_i) - \frac{e^2 A_L^2}{2}}{(p_i k_L)^2}. \tag{2.31b}$$

The latter line is obtained by a direct integration of the former analogous to Eq. (2.28). The results of [11, 12], in contrast to Eqs. (2.31a) and (2.31b), are given in a chosen reference frame but their generalization to a covariant form is straightforward and the results agree with Eqs. (2.31a) and (2.31b), as it must be.

To find now the classical emission formula for an electron moving in a plane wave field, these solutions are inserted into Eq. (2.17). To simplify the exponential phase, one writes it in the form

$$k_1 r(t) = \int_{-\infty}^{t} dt' k_1^\mu \frac{u_\mu}{\gamma(t')} = \int_{-\infty}^{\eta} d\eta' \frac{dt'}{d\eta'} k_1^\mu \frac{p_\mu}{m\gamma(t')}. \tag{2.32}$$

Now one may use the relation between the invariant phase of the incident laser pulse and the laboratory time

$$\frac{d\eta}{dt} = k_L^\mu \frac{dr_\mu(t)}{dt} = k_L^\mu \frac{p_\mu(t)}{m\gamma(t)}. \tag{2.33}$$

The numerator in this expression is given by the constant of motion $k_L^+ p_i^-$. Inserting furthermore Eqs. (2.31a) and (2.31b) we find

$$\begin{aligned} k_1 r(t) &= \frac{k_1^\mu}{p_i k_L} \int_{-\infty}^{\eta} d\eta' p_\mu(\eta') \\ &= \int_{-\infty}^{\eta} d\eta' \frac{k_1 p_i}{p_i k_L} - e\frac{k_1 A_L}{p_i k_L} \psi_A(\eta') - \frac{e^2 A_L^2 (k_1 k_L)}{2(p_i k_L)^2} \psi_A^2(\eta'). \end{aligned} \tag{2.34}$$

Finally, since it is customary to formulate all quantities in the interaction with a plane wave as functions of the wave's invariant phase η or equally the light-cone coordinate x^-, by virtue of Eq. (2.33) we reformulate Eq. (2.17) to yield

$$\frac{dE}{d\omega d\Omega} = \frac{e^2\omega^2}{4\pi^2 {p_i^-}^2} \left| \int_{-\infty}^{\infty} dx^- p^\mu(x^-) e^{ik_1 r(x^-)} \right|^2. \tag{2.35}$$

2.1.2 Interaction with a Monochromatic Plane Laser Wave

If the temporal duration of a laser pulse τ_L is much larger than its cycle period ω_L^{-1}, its spectrum will be very narrowly confined around ω_L. It is then a good approximation to model the spectrum as monochromatic, i.e. by a δ-spike in frequency space. The temporal structure of this type of fields is strictly periodic and allows for significant simplifications in the calculations. For instance, the trajectory of an electron inside a monochromatic plane wave field, reduces to a simple form. Observing it in a reference frame, in which the electron is on average at rest, it moves on a trajectory with a well-known *figure-8* shape [11, 12]. Due to this strictly monochromatic motion, the computation of the energy spectra can be largely simplified and central features of a QED computation of the scattering from a monochromatic laser field are already found in this classical analysis.

The following discussion largely follows [15] and summarizes the results of that work. For reasons of convenience and without loss of generality, however, we will analyze the interaction in a reference frame where the laser wave propagates along the positive z-direction ($\boldsymbol{k}_L = \omega_L(0, 0, 1)$) and the electron's energy and velocity in the absence of the field reduce to the free values ε_i and $\boldsymbol{\beta}_i = \beta_i(0, 0, -1)$, respectively. It was shown that in such a reference frame for a monochromatic laser field of the form $A^\mu_L(\eta) = A_L \epsilon^\mu_L \cos(\eta)$ Eq. (2.17) can be written as

$$\frac{dE^2}{d\omega d\Omega} = \frac{e^2\omega^2}{4\pi^2}\left[(1-n_x^2)K_x^2 - 2n_x n_y K_x K_y + (1-n_z^2)K_z^2\right] \tag{2.36}$$

where n_i are the components of the vector $\boldsymbol{n}$ pointing in the observation direction and it was defined the vector function

$$\boldsymbol{K} = \int_{-\infty}^{\infty} d\eta \frac{d\boldsymbol{r}(\eta)}{d\eta} \exp\left[\mathrm{i}\frac{\omega}{\omega_L}\left(\eta + \omega_L(z - \boldsymbol{n}\boldsymbol{r}(\eta))\right)\right]. \tag{2.37}$$

For the assumed monochromatic potential the trajectory, given by the spatial components of Eq. (2.31b), is found to be

$$\boldsymbol{r}(\eta) = \boldsymbol{a}\eta + \boldsymbol{b}\sin(\eta) + \boldsymbol{c}\sin(2\eta), \tag{2.38}$$

where the following constant vectors are defined

$$\boldsymbol{a} = \frac{1}{\omega_L}\left[\left(\frac{m\xi}{2\varepsilon_i}\right)^2 \frac{\boldsymbol{n}_L}{(1-\boldsymbol{n}_L\boldsymbol{\beta}_i)^2} + \frac{\boldsymbol{\beta}_i}{(1-\boldsymbol{n}_L\boldsymbol{\beta}_i)}\right] \tag{2.39a}$$

$$\boldsymbol{b} = \boldsymbol{\epsilon}_L \frac{\frac{m\xi}{\varepsilon_i}}{\omega_L(1-\boldsymbol{n}_L\boldsymbol{\beta}_i)} \tag{2.39b}$$

$$\boldsymbol{c} = \boldsymbol{k}_L \frac{\left(\frac{m\xi}{2\varepsilon_i}\right)^2}{2(\omega_L(1-\boldsymbol{n}_L\boldsymbol{\beta}_i))^2}. \tag{2.39c}$$

The exponential factors in Eq. (2.37) containing trigonometric functions can then be transformed to Bessel functions of integer order J_n via their generating function

$$\mathrm{e}^{\mathrm{i}u\sin(\eta)} = \sum_{n=-\infty}^{\infty} J_n(u)\mathrm{e}^{\mathrm{i}n\eta}. \tag{2.40}$$

This relation has been widely employed in problems involving intense monochromatic laser fields [11] and also takes an important role in respective QED analyses to assign photon numbers to an actually unquantized laser field [16].

The resulting expression of the integrals $\boldsymbol{K}$ then allows to perform the integration in η, yielding δ-functions of the form

$$\boldsymbol{K} \propto \sum_{n=-\infty}^{\infty} \ldots \delta \left(\omega - n \frac{\omega_L}{1 - \boldsymbol{a}(\boldsymbol{n} - \boldsymbol{n}_L)} \right). \tag{2.41}$$

The explicit form of the integrals is rather involved [15] but is not needed here. Taking the square of this δ-function, as is required for evaluating Eq. (2.36), by usual methods employed in S-Matrix calculations one obtains an expression for the overall emitted power in the form

$$\frac{dP}{d\Omega} \propto \sum_{n=0}^{\infty} \frac{dP^{(n)}}{d\Omega}. \tag{2.42}$$

The nth term in this series represents the power emitted into the nth harmonic, whose frequencies according to the conservation law of Eq. (2.41) are equidistantly distributed and given by

$$\omega^n = n \frac{\omega_L}{1 - \omega_L \boldsymbol{a}(\boldsymbol{n} - \boldsymbol{n}_L)}. \tag{2.43}$$

This result is not equivalent to Eq. (1.8), because of the radiation pressure of the laser field, which may reduce the Doppler shift. In fact, we note that in the limit $\xi \to 0$ Eq. (2.43) for $n = 1$ goes over to the expression

$$\omega = \omega_L \frac{(1 - \boldsymbol{n}_L \boldsymbol{\beta}_i)}{(1 - \boldsymbol{n}_L \boldsymbol{\beta}_i) - \boldsymbol{\beta}_i(\boldsymbol{n} - \boldsymbol{n}_L)} = \omega_L \frac{1 - \boldsymbol{n}_L \boldsymbol{\beta}_i}{1 - \boldsymbol{n} \boldsymbol{\beta}_i}, \tag{2.44}$$

which in fact is equivalent to the ordinary Doppler shift of Eq. (1.8). This result of the scattered harmonic frequencies alongside the utilization of the generating function of the Bessel functions according to Eq. (2.40) will enable us to compare the presented classical calculations to the QED analyses for monochromatic laser waves (see Eq. (2.40)).

2.1.3 Interaction with a Focused Laser

The assumption of plane wave, though often justified to a good extent, is never complete [17]. This is easily seen by recalling that any function $f(t - x^{\parallel})$, where $x^{\parallel}$ is the propagation direction of the laser, is a plane wave solution of the wave equation Eq. (2.3). This solution, however, does not feature a dependence on the two transversal coordinates $\boldsymbol{x}^{\perp}$. Consequently, at every phase value $\eta = k_L x$ the field is constant in a plane stretching infinitely in the $\boldsymbol{x}^{\perp}$-plane (hence the label *plane wave*).

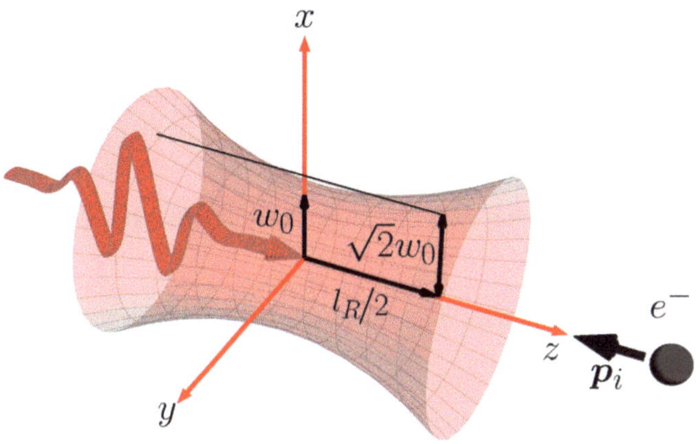

Fig. 2.3 Gaussian beam focus in a coordinate system according to Fig. 2.1

Though this is of course, rigorously speaking, unphysical, in many cases it is still a valid approximation, as we will see. To treat a laser beam consistently, however, it would much rather be necessary to include the natural transversal extent of the beam. The standard formalism of describing the field distribution of a monochromatic laser wave focused to a perpendicular spot size w_0 (see Fig. 2.3) is due to Davis [18]. Once analytic expressions for the laser fields are found, it is a straightforward task in classical electrodynamics to obtain the electron's trajectory from Eq. (2.11) and thus via Eq. (2.17) its emission pattern. We wish to briefly sketch the concept of this treatment. Assume the laser pulse's vector potential to be given by

$$A_L^\mu(x) = A_L \epsilon_L^\mu \Psi_L(\boldsymbol{x}) \mathrm{e}^{-\mathrm{i}k_L x} + \varepsilon_\phi^\mu \phi(x). \tag{2.45}$$

In this expression the space dependent factor $\Psi_L(\boldsymbol{r})$ is introduced to describe the spatial focusing of the laser pulse, $\epsilon_L^\mu = (0, \boldsymbol{\epsilon}_L)$ is the well known polarization vector of the laser pulse and $\varepsilon_\phi = (1, 0, 0, 0)$ denotes that the second term introduces a nontrivial scalar potential into $A_L^\mu(x)$. One can no longer assume the vector potential to be purely spatially polarized, since the Lorenz gauge condition could not be fulfilled in that case. To circumvent this difficulty it is customary to incorporate a nonzero scalar potential, as indicated in Eq. (2.45). The introduced scalar potential is connected with the spatial components of $A_L^\mu(x)$ via the Lorenz gauge condition

$$\partial_t \phi(x) = \nabla \Big(A_L \boldsymbol{\epsilon}_L \Psi_L(\boldsymbol{x}) \mathrm{e}^{-\mathrm{i}k_L x} \Big). \tag{2.46}$$

Thus, the four potential still obeys the Lorenz gauge condition $\partial_\mu A_L^\mu(x) = 0$. Due to Eq. (2.46) the problem is fully determined if we find a solution of the wave equation for the spatial vector potential $\boldsymbol{A}_L(x)$. We will thus restrict the following discussion to this quantity. The electric and magnetic fields, derived from Eq. (2.1), then of course also are no longer linearly polarized, but will exhibit longitudinal field components, typical of a focused beam. Inserting Eq. (2.45), the Lorenz gauge wave equation $\Box \boldsymbol{A}_L(x) = 0$ reduces to

$$\nabla^2 \Psi_L(\boldsymbol{x}) - 2\mathrm{i}\omega_L \frac{\partial \Psi_L(\boldsymbol{x})}{\partial x^{\|}} = 0. \tag{2.47}$$

Solving Eq. (2.47) exactly for $\Psi_L(\boldsymbol{x})$ would mean to find the class of electromagnetic potentials, that vary in time like $\mathrm{e}^{-\mathrm{i}\omega_L t}$ and fulfill the wave Eq. (2.3). Unfortunately such a complete solution has not been reported up to today, whence a perturbative ansatz for Eq. (2.47) is called for. Such an ansatz is found from the consideration that a laser beam cannot be focussed to spot sizes w_0 smaller than its central wavelength $\lambda_L = 2\pi/\omega_L < w_0$. A dimensionless and always small parameter, characterizing the focusing of the laser beam, is then given by

$$s_L = \frac{1}{\omega_L w_0} = \frac{\lambda_L}{2\pi w_0}. \tag{2.48}$$

In addition to this perpendicular confinement, a laser focus also exhibits a characteristic longitudinal spreading length $l_R = w_0/s_L = \omega_L w_0^2$, often called *Rayleigh length.* At a distance $z = l_R/2$ from the focal plane the laser's intensity has dropped to half its value at $z = 0$, whence l_R is often referred to as the longitudinal extent of the focal spot. It is then useful to transform Eq. (2.47) to the dimensionless variables $\boldsymbol{\rho} = (\rho^{\|}, \rho_1^{\perp}, \rho_2^{\perp})$ with $\rho^{\|} = x^{\|}/l_R$ and $\boldsymbol{\rho}^{\perp} = \boldsymbol{x}^{\perp}/w_0$, resulting in

$$\left(\frac{\partial^2}{\partial^2 \rho_1^{\perp}} + \frac{\partial^2}{\partial^2 \rho_2^{\perp}}\right)\Psi_L(\boldsymbol{\rho}) + s_L^2 \frac{\partial^2 \Psi_L}{\partial {\rho^{\|}}^2} - 2\mathrm{i}\frac{\partial \Psi_L(\boldsymbol{\rho})}{\partial \rho^{\|}} = 0. \tag{2.49}$$

From this equation we can guess the correct way of a perturbative ansatz for the solution. In fact, assuming that the focus parameter would vanish ($s_L = 0$), Eq. (2.49) is solved by the function [18]

$$\Psi_L^0(\boldsymbol{\rho}) = \left(\frac{w_0^2}{w^2(\rho^{\|})} + \mathrm{i}\frac{l_R}{2R(\rho^{\|})}\right)\exp\left[-\left(\frac{w_0^2}{w^2(\rho^{\|})} + \mathrm{i}\frac{l_R}{2R(\rho^{\|})}\right)\rho_{\perp}^2\right], \tag{2.50}$$

where the following definitions are used

$$w(\rho^{\|}) = w_0\sqrt{1 + 4{\rho^{\|}}^2} \tag{2.51a}$$

$$R(\rho^{\|}) = l_R \rho^{\|}\left(1 + \frac{1}{4{\rho^{\|}}^2}\right). \tag{2.51b}$$

From Eq. (2.50) we read off that $w(\rho^{\|})$ gives the perpendicular extent of the laser focus in dependence of the longitudinal position (note that at $|\boldsymbol{\rho}^{\perp}| = w(\rho^{\|})$ the function $\Psi_L^0(\boldsymbol{\rho})$ is always damped by at least a factor e^{-1}). The factor $R(\rho^{\|})$ gives the radius of curvature of the non-plane wavefront going through the laser-axis at $x^{\|} = l_R \rho^{\|}$. The meaning of the quantity l_R is also apparent. Its half (corresponding to $\rho^{\|} = 1/2$) indicates the distance from the origin $\rho^{\|} = 0$ along the laser's propagation direction, after which the focus' perpendicular extent has increased to $w(\rho^{\|} = 1/2) = \sqrt{2}w_0$, as indicated in Fig. 2.3. In this sense l_R can be interpreted as a measure of the overall longitudinal extent of the laser focus. Having now found a lowest order perturbative solution of Eq. (2.49) we can readily guess the proper form of a perturbative expansion of the complete focusing function to be [18]

$$\Psi_L(\rho) = \sum_{n=0}^{\infty} s_L^{2n} \Psi_L^{2n}(\rho). \tag{2.52}$$

Inserting this ansatz into Eq. (2.49) we find as determining equation for the second term in this series

$$\left(\frac{\partial^2}{\partial^2 \rho_1^{\perp}} + \frac{\partial^2}{\partial^2 \rho_2^{\perp}} - 2\mathrm{i}\frac{\partial}{\partial \rho^{\|}} \right) \Psi_L^2(\rho) = -\frac{\partial^2 \Psi_L^0(\rho)}{\partial {\rho^{\|}}^2}, \tag{2.53}$$

and all higher orders accordingly. From the thusly found vector potential Eq. (2.45) it is simple to derive the electromagnetic field strength tensor $F_L^{\mu\nu}(x)$ and hence, via e.g. numerical integration of Eq. (2.11), an electron's trajectory in the focused laser beam (compare Fig. 2.2). The procedure just outlined, however, is not applicable for pulsed laser fields, since it is found assuming a time variation of the form $\mathrm{e}^{-\mathrm{i}\omega_L t}$ for all times $t \in [-\infty, \infty]$ with a fixed laser frequency ω_L. To describe a focused laser pulse the ansatz for the spatial vector potential of Eq. (2.45) can be modified according to [17]

$$\boldsymbol{A}_L(x) = g(\eta) \boldsymbol{\epsilon}_L \Psi_L(\boldsymbol{x}) \mathrm{e}^{-\mathrm{i}k_L x}, \tag{2.54}$$

where the function $g(\eta)$ is used to introduce an arbitrary temporal shaping of the laser pulse, depending only on the invariant phase $\eta = \omega_L(t - x^{\|})$. The determining equation for the focusing function then turns into

$$\nabla^2 \Psi_L(\boldsymbol{x}) - 2\mathrm{i}\omega_L \frac{\partial \Psi_L(\boldsymbol{x})}{\partial x^{\|}} \left(1 - \mathrm{i}\frac{\partial_\eta g(\eta)}{g(\eta)} \right) = 0. \tag{2.55}$$

It is then usual to assume, that the envelope function's derivative is small compared to its function value $\partial_\eta g(\eta) \ll g(\eta)$, which is a valid approximation for a long laser pulse. This is also called *slowly varying envelope approximation* [19, 20]. For the case of few-cycle laser pulses, however, the approximation of a slowly varying envelope is not a good one. In fact, for few-cycle laser pulses, which are well described by the model Eq. (2.6) we find the assumedly small fraction to be $\partial_\eta g(\eta)/g(\eta) = -(2\cot(\eta/2n_C))/n_C \propto n_C^{-1}$. For $n_C \sim 1$, corresponding to a few-cycle laser pulse, this quantity is non-negligible, and for $\eta \to 2\pi n_C$ it even diverges. Consequently, for few-cycle pulses a new approximation scheme is called for, to describe a focused laser pulse.

To properly account for spatial focusing of a few-cycle laser pulse we propose a scheme which is based on the idea to impose a spatial focusing on a previously plane wave laser potential. Before doing so, however, we suggest to decompose the plane wave potential into its Fourier components and then consider the focusing of each separate (monochromatic) mode to the same focal spot size. In spirit of this idea, we note that any shape function which vanishes outside of a finite interval $[0, \tau_L]$ has a finite Fourier sum according to

$$\psi_{\mathcal{A}}(\eta) = \sum_{n=-N}^{N} c_n \mathrm{e}^{\mathrm{i}n\,\omega_0 (t - x^{\parallel})} \tag{2.56}$$

with the fundamental frequency $\omega_0 = {}^{2\pi\omega_L}/_{\tau_L}$. The factors c_n are the usual Fourier series coefficients

$$c_n = \frac{1}{\tau_L} \int_0^{\tau_L} d\eta\; \psi_{\mathcal{A}}(\eta) \mathrm{e}^{-\mathrm{i}n\omega_0 \eta/\omega_L}. \tag{2.57}$$

From here on we will outline the proposed method for the specific model of a few-cycle pulse Eq. (2.6) and we note that in this case it is $\tau_L = 2\pi n_C$ and we thus find the simple relation $\omega_0 = {}^{\omega_L}/_{n_C}$. Employing Eq. (2.56) instead of Eq. (2.54), the ansatz for the vector potential becomes

$$\boldsymbol{A}_L(x) = \epsilon_L \sum_{n=-N}^{N} c_n \Psi_{L,n}(\boldsymbol{x}) \mathrm{e}^{\mathrm{i}n\,\omega_0 (t - x^{\parallel})}. \tag{2.58}$$

Plugging this expression into the wave equation we arrive at a differential equation for the determination of the $\Psi_{L,n}(\boldsymbol{x})$ analogous to Eq. (2.47)

$$\sum_{n=-N}^{N} c_n \left(\nabla^2 \Psi_{L,n}(\boldsymbol{x}) - 2\mathrm{i} \left(n \frac{\omega_L}{n_C} \right) \frac{\partial \Psi_{L,n}(\boldsymbol{x})}{\partial x^{\parallel}} \right) = 0. \tag{2.59}$$

Since the separate frequency components represented by the single terms of this series do not mix, a solution of the above equation can be readily written down in terms of solutions of Eq. (2.47). We accordingly find that the focusing functions for the Fourier modes of the vector potential Eq. (2.58) are given by

$$\Psi^0_{L,n}(\boldsymbol{\rho_n}) = \left(\frac{w_0^2}{w_n^2(\rho_n^{\parallel})} + \mathrm{i} \frac{l_{R,n}}{2R_n(\rho_n^{\parallel})} \right) \exp\left[- \left(\frac{w_0^2}{w_n^2(\rho_n^{\parallel})} + \mathrm{i} \frac{l_{R,n}}{2R_n(\rho_n^{\parallel})} \right) \rho_\perp^2 \right]. \tag{2.60}$$

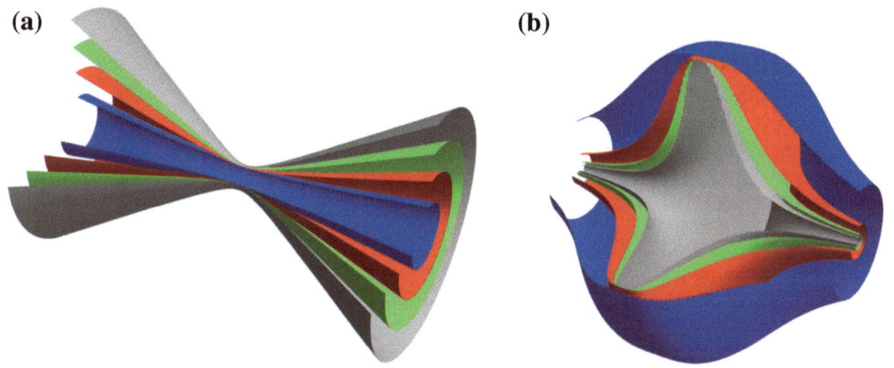

Fig. 2.4 Fourier frequency component foci of a pulse derived from inserting Eq. (2.6) into Eq. (2.56) for $n_C = 2$. The only non-vanishing Fourier components are $n = 1$ (*blue*), $n = 2$ (*red*), $n = 3$ (*green*), $n = 4$ (*gray*). **a** Surfaces of constant $w(\rho^{\|})$, all coinciding with w_0 at $\rho^{\|} = 0$. **b** Surfaces of constant intensity, all featuring the same peak intensity

It is now essential that for all n the same w_0 enters the equation, corresponding to a focusing to the same spot size. The definitions of the remaining variables are analogous to the presented monochromatic analysis

$$
\begin{aligned}
l_{R,n} &= n\frac{\omega_L}{n_C}w_0^2 \\
\rho_n^{\|} &= \frac{x^{\|}}{l_{R,n}} \\
w_n(\rho_n^{\|}) &= w_0\sqrt{1+4{\rho_n^{\|}}^2} \\
R_n(\rho_n^{\|}) &= l_{R,n}\rho_n^{\|}\left(1+\frac{1}{4{\rho_n^{\|}}^2}\right).
\end{aligned}
$$

The computation of higher order terms is carried out according to Eq. (2.53). Essentially the found behaviour is explained by the observation that the focusing to a Gaussian beam is a linear operation on the laser field and doesn't mix its frequency components. Accordingly one can picture the focused laser beam as a superposition of frequency components focused to the same focal spot as sketched in Fig. 2.4. We wish to point out that Eqs. (2.58) and (2.60) are exact in the temporal focusing of the plane wave field. There was no need for a slowly varying envelope approximation $\partial_\eta g(\eta)/g(\eta) \ll 1$. In such a short laser pulse, focussed to small spot sizes, the classical electron trajectory may be significantly changed, as can be traced in Fig. 2.5, where we show the trajectory only within the ϵ_L-$\boldsymbol{k}_L$ plane, as was motivated below Fig. 2.2. From that figure we also conclude that the plane wave approximation gives useful results for the trajectory already for a focusing as small as $w_0 \geq 2\lambda_L$.

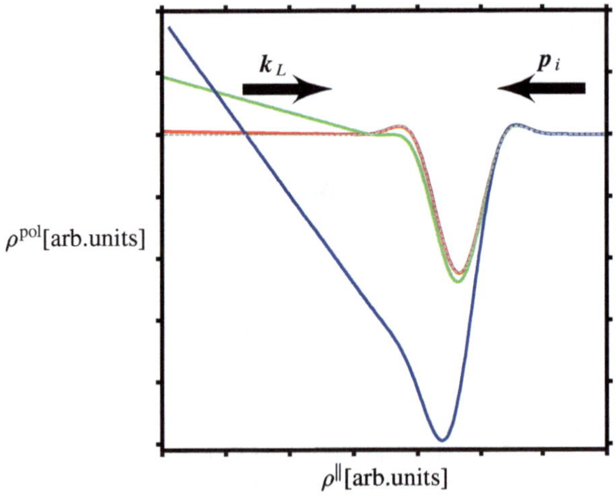

Fig. 2.5 Trajectories of an electron colliding head on with a laser pulse with shape derived from Eq. (2.6) with $n_C = 2$ and central wavelength $\lambda_L = 800\,\text{nm}$ for a focusing of $w_0 = 2\lambda_L$ (*red*), $w_0 = \lambda_L$ (*green*) and $w_0 = \lambda_L/2$ (*blue*, corresponds to a focal spot size of λ_L). For comparison the plane wave trajectory is shown in *gray dashes*

2.2 Quantum Electrodynamics

It is in order to shortly review the fundamental concepts of QED. The basic concept of this highly successful theory is, that all electrically charged elementary particles, as well as the gauge particles mediating the electromagnetic interaction between them, have to be interpreted as excitations of quantum fields. The fundamental quantity of the accordingly required field theory for the coupling of a charged spinor field $\Psi(x)$ with an electromagnetic potential $A^\mu(x)$ is the Lagrangian of quantum electrodynamics [21, 22]. The Lagrangian is not unique, since due to gauge invariance several expressions will lead to equivalent equations of motion of the described fields. One may take advantage of this freedom and employ an expression of the QED Lagrangian, which circumvents several problems such as the vanishing of the conjugate momentum of the photon field [23, 24]. To enable a canonical quantization scheme for the photon field, we will employ the so-called Fermi-Lagrangian [25] for its description, turning the QED Lagrangian into the expression

$$\mathcal{L}_{\rm QED} = \overline{\Psi}(x)\,(\mathrm{i}\not{\mathcal{D}} - m)\,\Psi(x) - \frac{1}{8\pi}\partial_\nu A_\mu(x)\partial^\mu A^\nu(x). \tag{2.61}$$

Here $\mathcal{D}^\mu(x) = \partial^\mu + \mathrm{i}eA^\mu(x)$ is the so-called gauge covariant derivative [22]. The equations of motion for the involved fields $\Phi(x) \in \left[\Psi(x), \overline{\Psi}(x), A^\mu(x)\right]$ are obtained as extremal points of the variation of the action with respect to the corresponding fields, as is usual in Lagrangian field theory, resulting in the Euler-Lagrange equations

$$\partial^\mu\left(\frac{\partial\mathcal{L}}{\partial(\partial^\mu\phi(x))}\right) - \frac{\partial\mathcal{L}}{\partial\phi(x)} = 0. \tag{2.62}$$

This procedure results in the Dirac equation and its conjugate for the spinor field, its conjugate and the photon field, respectively [26]

$$(\mathrm{i}\not{D}(x) - m)\Psi(x) = 0 \tag{2.63a}$$

$$\overline{\Psi}(x)(\mathrm{i}\not{D}(x) + m) = 0 \tag{2.63b}$$

$$\Box A^{\mu}(x) = 4\pi e j^{\mu}_{\mathrm{Dirac}}(x) = 4\pi e\overline{\Psi}(x)\gamma^{\mu}\Psi(x). \tag{2.63c}$$

Given in this form the fields $\Psi(x), \overline{\Psi}(x), A^{\mu}(x)$ are classical unquantized quantities. Since one is now interested in a relativistic theory of interacting particles, one naturally has to consider a multi-particle theory, since particles may be created or annihilated taking or providing the amount of energy corresponding to their rest masses, respectively [22]. The fields will then be represented by operators, acting on a space of physical quantum states, representing the number of existing particles. To quantize the theory one introduces conjugate momenta for the fields according to

$$\pi_{\Psi}(x) = \frac{\partial\mathcal{L}}{\partial(\partial_t\Psi(x))} = \mathrm{i}\overline{\Psi}(x)\gamma^0 = \mathrm{i}\Psi^{\dagger}(x) \tag{2.64a}$$

$$\pi_{\overline{\Psi}}(x) = \frac{\partial\mathcal{L}}{\partial(\partial_t\overline{\Psi}(x))} = 0 \tag{2.64b}$$

$$\pi^{\mu}_{A}(x) = \frac{\partial\mathcal{L}}{\partial(\partial_t A_{\mu}(x))} = -\frac{1}{4\pi}\partial_t A^{\mu}(x). \tag{2.64c}$$

The problem that in these equations the conjugate momentum of the conjugate spinor field $\overline{\Psi}$ vanishes, can be resolved by symmetrizing the Lagrangian (2.61) and does not cause any particular difficulties [21]. The Hamiltonian of the theory is formed in the standard way

$$\begin{aligned} H(t) &= \int \mathrm{d}\boldsymbol{x}\, \left(\pi_{\Psi}(x)\partial_t\Psi(x) + \pi^{\mu}_{A}(x)\partial_t A_{\mu}(x) - \mathcal{L}(\Psi(x), \overline{\Psi}(x), A^{\mu}(x))\right) \\ &= H_{\mathrm{Dirac}}(t) + H_{\mathrm{Maxwell}}(t) + H_{\mathrm{int}}(t), \end{aligned} \tag{2.65}$$

$$H_{\mathrm{Dirac}}(t) = \int d\boldsymbol{x}\, \overline{\Psi}(x)(-\mathrm{i}\gamma\nabla + m)\Psi(x)$$

$$H_{\mathrm{Maxwell}}(t) = \frac{1}{8\pi}\int d\boldsymbol{x}\, \left(-\pi^{\mu}_{A}(x)\pi_{A,\mu}(x) + \partial_{\nu}A_{\mu}(x)\partial^{\mu}A^{\nu}(x)\right)$$

$$H_{\mathrm{int}}(t) = e\int d\boldsymbol{x}\, j^{\mu}_{\mathrm{Dirac}}(x)A_{\mu}(x).$$

To now quantize the theory one interprets the above fields, as well as the derived Hamiltonian, as Schrödinger picture operators, indicated by a superscript S, and subjects them to the anti-commutation and commutation relations for the fermionic and bosonic fields, respectively

$$\left[\hat{\Psi}_r^S(\boldsymbol{x}),\, \hat{\Psi}_q^{S\dagger}(\boldsymbol{x}')\right]_+ = \delta_{rq}\delta^{(3)}(\boldsymbol{x}-\boldsymbol{x}') \tag{2.66a}$$

$$\left[\hat{A}^{S\mu}(\boldsymbol{x}),\, \hat{\pi}_A^{S\nu}(\boldsymbol{x}')\right]_- = \mathrm{i}g^{\mu\nu}\delta^{(3)}(\boldsymbol{x}-\boldsymbol{x}') \tag{2.66b}$$

$$\left[\hat{\Psi}_r^S(\boldsymbol{x}),\, \hat{\Psi}_q(\boldsymbol{x}')\right]_+ = \left[\hat{\Psi}_r^{S\dagger}(\boldsymbol{x}),\, \hat{\Psi}_q^{S\dagger}(\boldsymbol{x}')\right]_+ = 0$$

$$\left[\hat{A}^{S\mu}(\boldsymbol{x}),\, \hat{A}^{S\nu}(\boldsymbol{x}')\right]_- = \left[\hat{\pi}_A^{S\mu}(\boldsymbol{x}),\, \hat{\pi}_A^{S\nu}(\boldsymbol{x}')\right]_- = 0.$$

where we have written the spin indices of the spinor fields $\hat{\Psi}^S$, $\hat{\overline{\Psi}}^S$ explicitly. In the Schrödinger picture operators do not carry a dynamic time dependency, whence the time dependency in Eq. (2.65) can be only due to an explicitly time dependent field, as e.g. an external electromagnetic current. The dynamical evolution of quantum systems is governed by the evolution of the quantum states, which obey the Schrödinger equation

$$\mathrm{i}\frac{d}{dt}\big|\Psi, A;\, (t)\rangle^S = \hat{H}^S\big|\Psi, A;\, (t)\rangle^S, \tag{2.67}$$

where $\hat{H}^S$ is the Schrödinger picture Hamiltonian from Eq. (2.65), with the entering fields understood as operators. For the sake of simplicity we do not consider an explicit time dependence of the Hamiltonian. Eq. (2.67) can formally be integrated by defining a time evolution operator

$$\big|\Psi, A;\, (t)\rangle^S = \hat{U}^S(t, t_0)\big|\Psi, A;\, (t_0)\rangle^S \tag{2.68}$$

$$\hat{U}^S(t, t_0) = \hat{\mathcal{T}}\exp\left[-\mathrm{i}\int_{t_0}^{t} dt\hat{H}^S\right], \tag{2.69}$$

where $\hat{\mathcal{T}}$ denotes time ordering [21, 25]. Even though we do not consider an explicitly time dependent Hamiltonian at this point, we introduce this general notion already for later convenience. For Eq. (2.69), however, no exact solution is known. It is thus customary to transform the quantum states and operators to the interaction picture of quantum dynamics. To this end one splits up the full Hamiltonian into its free and its interaction part

$$\hat{H}^S = \hat{H}_0^S + \hat{H}_{\text{int}}^S \tag{2.70a}$$

$$\hat{H}_0^S = \hat{H}_{\text{Dirac}}^S + \hat{H}_{\text{Maxwell}}^S. \tag{2.70b}$$

The transition from the Schrödinger to the interaction picture is accomplished by the unitary transformation

$$\hat{\mathcal{O}}^I(t) = \hat{U}_0^{S\dagger}(t, t_0)\hat{\mathcal{O}}^S\hat{U}_0^S(t, t_0) \tag{2.71a}$$

$$\big|\Psi, A;\, (t)\big\rangle^I = \hat{U}_0^{S\dagger}(t, t_0)\big|\Psi, A;\, (t)\big\rangle^S \tag{2.71b}$$

$$\hat{U}_0^S(t, t_0) = \hat{\mathcal{T}}\exp\left[-\mathrm{i}\int_{t_0}^{t} dt\hat{H}_0^S\right], \tag{2.71c}$$

where the superscript I denotes states and operators in the interaction picture. In this picture both states and operators carry a dynamic time evolution, however, with different governing equations. For the evolution of states in the interaction picture, we find a Schrödinger-type, whereas the time dependence of the operators is determined by a Heisenberg-type equation

$$\mathrm{i}\frac{d}{dt}\hat{\mathcal{O}}^I(t) = \left[\hat{\mathcal{O}}^I, \hat{H}_0^I\right] \tag{2.72a}$$

$$\mathrm{i}\frac{d}{dt}\big|\Psi, A;\, (t)\big\rangle^I = \hat{H}_{\mathrm{int}}^I\big|\Psi, A;\, (t)\big\rangle^I. \tag{2.72b}$$

We note that due to the transformation law (2.71a) it holds $\hat{H}_0^I \equiv \hat{H}_0^S$. Evaluating Eq. (2.72a) for the field operators $\hat{\Psi}(x)$ and $\hat{A}^\mu(x)$ we recover wave equations formally equivalent to the classical equations of motion for the field modes Eqs. (2.63). To describe all degrees of freedom of the theory, we have to expand the full field operator in a complete basis of solutions of the wave equation with operator-valued coefficients. We have to pay attention that the wave Eqs. (2.63) allow positive and negative energy solutions for the photon and the spinor fields (see Appendix B). Both contributions have to be included in a complete basis, whence the expanded field operators read

$$\hat{A}^\mu(x) = \int d\boldsymbol{k}\left[\hat{a}_{\boldsymbol{k}}A_{\boldsymbol{k}}^\mu(x) + \hat{a}_{\boldsymbol{k}}^\dagger A_{\boldsymbol{k}}^{*\mu}(x)\right] \tag{2.73a}$$

$$\hat{\Psi}(x) = \int d\boldsymbol{p}\left[\hat{c}_{\boldsymbol{p}}\Psi_p(x) + \hat{d}_{\boldsymbol{p}}^\dagger\Psi_{-p}(x)\right] \tag{2.73b}$$

$$\hat{\overline{\Psi}}(x) = \int d\boldsymbol{p}\left[\hat{c}_{\boldsymbol{p}}^\dagger\overline{\Psi}_p(x) + \hat{d}_{\boldsymbol{p}}\overline{\Psi}_{-p}(x)\right]. \tag{2.73c}$$

In this expression $A_{\boldsymbol{k}}^\mu(x)$ are the solutions of the photon field wave equation to the spatial wave vector $\boldsymbol{k}$ and $\hat{a}_{\boldsymbol{k}}$, $\hat{a}_{\boldsymbol{k}}^\dagger$ are creation and annihilation operators of this field mode. In the fermionic field operators $\Psi_{\pm p}(x)$ and $\overline{\Psi}_{\pm p}(x)$ are positive and negative energy solutions of the free wave equation for the fermionic fields to the spatial momenta $\pm\boldsymbol{p}$ and the coefficients $\hat{c}_{\boldsymbol{p}}$ and $\hat{c}_{\boldsymbol{p}}^\dagger$ ($\hat{d}_{\boldsymbol{p}}, \hat{d}_{\boldsymbol{p}}^\dagger$) are the according electronic (positronic) creation and annihilation operators, respectively.

The Schrödinger-type Eq. (2.72b), governing the time evolution of quantum states in the interaction picture, can be formally integrated out, in analogy to Eq. (2.68), yielding an explicit time evolution of the form

$$\big|\Psi, A; (t)\rangle^I = \hat{U}^I(t, t_0)\big|\Psi, A; (t_0)\rangle^I \tag{2.74}$$
$$\hat{U}^I(t, t_0) = \hat{\mathcal{T}} \exp\left[-\mathrm{i}\int\limits_{t_0}^{t} dt\hat{H}^I_{\mathrm{int}}\right],$$

Although also for Eq. (2.74) no closed analytical solution has been found so far, this type of time evolution is well suited for a perturbative approach, due to the smallness of the coupling constant $|e| \approx 137^{-1/2}$ at low energies (recall $H_{\mathrm{int}} \propto e$). In fact, it is possible to truncate the exponential series, as which the operator $\hat{U}^I(t, t_0)$ is defined, at a desired order of accuracy of perturbation theory. Every interaction between the fermionic and photon fields is then treated as a small perturbation of the free theories. The result of this procedure are the well-known Feynman diagrams of free QED (see Sect. 2.2.2 and [21, 22]). As a fixed basis for the space of states evolving according to Eq. (2.74), one chooses the Fock representation, with each state $\big|\ldots n_{\boldsymbol{k}} \ldots; \ldots n_{\boldsymbol{p}} \ldots\rangle$ corresponding to a given number of photons $n_{\boldsymbol{k}}$ in the momentum mode $\boldsymbol{k}$ and a number of massive fermions $n_{\boldsymbol{p}}$ in the mode $\boldsymbol{p}$. The ground state of this basis is then the vacuum state $\big|0\rangle$ defined by the action of the field annihilators on it

$$\hat{a}_{\boldsymbol{k}}\big|0\rangle = \hat{c}_{\boldsymbol{p}}\langle|0\rangle = \hat{d}_{\boldsymbol{p}}\big|0\rangle = 0. \tag{2.75}$$

A particular Fock state representing l $\boldsymbol{k}$-mode photons and $m(n)$ electrons with momentum $\boldsymbol{p}_{e^-}$ (positrons with momentum $\boldsymbol{p}_{e^+}$), is then formed from this vacuum state by the action of field construction operators on it

$$|l^{\gamma}_{\boldsymbol{k}} \ldots; \ldots m_{\boldsymbol{p}_{e^-}} \ldots; \ldots n_{\boldsymbol{p}_{e^+}}\rangle \propto \left(\hat{a}^{\dagger}_{\boldsymbol{k}}\right)^l \ldots \left(\hat{c}^{\dagger}_{\boldsymbol{p}_{e^-}}\right)^m \ldots \left(\hat{d}^{\dagger}_{\boldsymbol{p}_{e^+}}\right)^n \ldots |0\rangle. \tag{2.76}$$

2.2.1 Quantization in the Presence of a Strong External Field

A perturbative expansion, as outlined in the last section is no longer possible, if the coupling between the spinor and the electromagnetic fields, mediated by the coupling term $-e\int dt\, j_\mu A^\mu$ in the Lagrange function (2.61), can no longer be treated as a small perturbation. As was pointed out in Chap. 1, the expansion parameter of the perturbation series of the interaction between an electron and several photons from a strong laser field is the intensity parameter ξ. Hence, a laser field exceeding $\xi \gtrsim 1$ cannot be accounted for perturbatively. Similarly, there are other strong fields imaginable, that do not lend themselves to a perturbative treatment. In this case the spinor field has to be quantized in the presence of the electromagnetic background field. This task is canonically reached by investigating the quantum dynamics in the so-called Furry picture of quantum dynamics [21, 24, 27]. We will see, however, that large bits of the simpler discussion of the free theory still maintain their validity with slightly changed definitions

The essential concept of the Furry picture is to employ a split up of the QED-Hamiltonian differing from Eq. (2.70). To this end, one takes advantage of the physical fact, that electromagnetic fields, which are sufficiently strong to render the perturbative approach disfavorable, usually fulfill two assumptions: All photons in the field stem from one coherent source, realized by e.g. an atomic nucleus or a laser field. Secondly due to the tremendous photon-flux densities present in high-intensity laser fields, these fields can be treated as unquantized, neglecting the single photons' quantum dynamics. To obtain the Furry picture one can then split up the electromagnetic potential entering the Hamiltonian Eq. (2.65) into two separate components

$$A^{\mu}(x) = A^{\mu}_{\text{ext}}(x) + A^{\mu}_{\text{rad}}(x), \tag{2.77}$$

where $A^{\mu}_{\text{ext}}(x)$ is the explicitly time dependent strong *external* electromagnetic potential. To treat the potential $A^{\mu}_{\text{ext}}(x)$ as an unquantized field, one must not raise it to an operator level and impose no commutation relations on its components. Hence any contribution to the Hamiltonian operator, depending solely on $A^{\mu}_{\text{ext}}(x)$ can be omitted. The second contribution $A^{\mu}_{\text{rad}}(x)$ is the total of all remaining electromagnetic field modes, not belonging to the strong external field. In particular all single emitted (or absorbed) photons are excitations of this field term, hence the index referring to *radiation*. Inserting this expansion into Eq. (2.65), we find that the QED Hamiltonian takes the form

$$H(t) = H_{\text{Dirac}}(t) + H^{\text{ext}}_{\text{int}}(t) + H^{\text{rad}}_{\text{Maxwell}}(t) + H^{\text{rad}}_{\text{int}}(t). \tag{2.78}$$

The term $H_{\text{Dirac}}(t)$ is the same as in Eq. (2.65), the terms $H^{\text{ext,rad}}_{\text{Maxwell}}(t)$ are derived from that equation by replacing $A^{\mu} \rightarrow A^{\mu}_{\text{ext,rad}}$. In Eq. (2.78) terms coupling the two four potentials A^{μ}_{ext} to A^{μ}_{rad} were already dropped, since they do not influence the equations of motion of these fields. In the Furry picture the above Hamiltonian is split up according to

$$H(t) = H^{\text{Furry}}_{0}(t) + H^{\text{Furry}}_{\text{int}}(t) \tag{2.79}$$

$$H^{\text{Furry}}_{0}(t) = H_{\text{Dirac}} + H^{\text{ext}}_{\text{int}}(t) + H^{\text{rad}}_{\text{Maxwell}}$$

$$H^{\text{Furry}}_{\text{int}} = H^{\text{rad}}_{\text{int}}, \tag{2.80}$$

where the explicit time dependence of the free Hamiltonian of the Furry picture $H^{\text{Furry}}_{0}(t)$ is indicated. The transition from the Schrödinger to the Furry picture is accomplished by the unitary transformation

$$\hat{O}^{F}(t) = \left(\hat{U}^{\text{Furry}}_{0}(t, t_0)\right)^{\dagger} \hat{O}^{S} \hat{U}^{\text{Furry}}_{0}(t, t_0) \tag{2.81a}$$

$$\big|\Psi, A; (t)\big\rangle^F = \left(\hat{U}_0^{\text{Furry}}(t, t_0)\right)^\dagger \big|\Psi, A; (t)\big\rangle \tag{2.81b}$$

$$\hat{U}_0^{\text{Furry}}(t, t_0) = \hat{\mathcal{T}} \exp\left[-\mathrm{i}\int_{t_0}^{t} dt \hat{H}_0^{\text{Furry}}(t)\right]. \tag{2.81c}$$

Due to the explicit time dependence of $\hat{H}_0^{\text{Furry}}(t)$, caused by the external electromagnetic current, the time ordering is indispensable in the definition of the time evolution operator of the Furry picture $\hat{U}_0^{\text{Furry}}(t, t_0)$. We can then largely adopt the discussion subsequent to Eq. (2.70), albeit, respecting the changed definitions of the free and interaction Hamiltonian. In particular, the quantization schemes of the fermionic fields and the field A^μ_{rad} are analogous to the discussion of the previous section. The intricacies of the fact that now the free Hamilton operator $\hat{H}_0^{\text{Furry}}(t)$ is explicitly time dependent can be found e.g. in [24]. We do not repeat those discussions but directly turn to discussing the dynamic evolutions of the quantized fermionic field operators. Analogous to Eq. (2.72) we find the dynamic evolution of operators and states in the Furry picture to be governed by the equations

$$\mathrm{i}\frac{d}{dt}\hat{O}^F(t) = \left[\hat{O}^F, \hat{H}_0^{\text{Furry}}\right] \tag{2.82a}$$

$$\mathrm{i}\frac{d}{dt}\big|\Psi, A; (t)\big\rangle^F = \hat{H}^F_{\text{int}}\big|\Psi, A; (t)\big\rangle^F. \tag{2.82b}$$

Please note that the interaction Hamiltonian in the Furry picture is given by the expression

$$\hat{H}^F_{\text{int}} = \left(\hat{U}_0^{\text{Furry}}(t, t_0)\right)^\dagger \hat{H}^{\text{Furry}}_{\text{int}} \hat{U}_0^{\text{Furry}}(t, t_0) = e\int d\boldsymbol{x}\, \hat{\overline{\Psi}}^F \hat{\not{A}}^F_{\text{rad}} \hat{\Psi}^F. \tag{2.83}$$

Solving the operator equation Eq. (2.82a), we find the wave equation

$$\left(\not{\partial} - e\hat{\not{A}}^F_{\text{ext}} - m\right)\hat{\Psi}^F(x) = 0, \tag{2.84}$$

which is just the Dirac equation in the presence of the assumedly strong external potential A^μ_{ext}. We find that the operators of the fermionic fields are given by an expression analogous to Eq. (2.73)

$$\hat{\Psi}_{A_{\text{ext}}}(x) = \int d\boldsymbol{p}\left[\hat{c}_{\boldsymbol{p}}\Psi_{p,A_{\text{ext}}}(x) + \hat{d}^\dagger_{\boldsymbol{p}}\Psi_{-p,A_{\text{ext}}}(x)\right] \tag{2.85a}$$

$$\hat{\overline{\Psi}}_{A_{\text{ext}}}(x) = \int d\boldsymbol{p}\left[\hat{c}^\dagger_{\boldsymbol{p}}\overline{\Psi}_{p,A_{\text{ext}}}(x) + \hat{d}_{\boldsymbol{p}}\overline{\Psi}_{-p,A_{\text{ext}}}(x)\right], \tag{2.85b}$$

where only in this case the wave functions $\Psi_{\pm p,A_{\text{ext}}}(x)$ need to fulfill the Dirac equation in the presence of A_{ext}, analogous to Eq. (2.84) and the coefficients $\hat{c}_p$ and $\hat{c}_p^{\dagger}$ ($\hat{d}_p$, $\hat{d}_p^{\dagger}$) are the creation and annihilation operators of these field modes.

Summarizing the above discussion we state that, to obtain the Furry picture, the potential term A^{μ}_{ext} is attributed to the free Hamiltonian H_0 of the interaction picture. Hence it becomes clear, why this contribution to the electromagnetic potential is labeled *external*: It enters the equation of motion of the spinor field as an additional term, independent of the interaction of the spinors with the radiation field. Furthermore, since A^{μ}_{ext} is not written as an operator term, it is obvious that in the Furry picture the external potential is not quantized, but treated as a classical current. The dynamic evolution of the Furry picture states, given by Eq. (2.82b), is determined by the radiation field modes. The interaction with these - assumedly weak - field modes, however, is again of order α_{QED} and thus accessible to a perturbative expansion as we wish to outline in the following section.

2.2.2 *The S-Matrix Expansion*

A key role in the investigation of QED is taken by scattering experiments. A typical experiment of this kind would be the respective scattering of an electron and an intense laser pulse. To describe such scenarios theoretically, one relies on the *S -Matrix formalism*, where the S can be understood to mean *scattering*. The idea underlying this formalism is, that the interaction is confined to a small region in space and time, as is reasonable for a realistic laboratory experiment. The particles entering and leaving the interaction region can then be considered to origin from and propagate to an infinitely remote past and future, respectively, without further interaction. In the Furry picture the evolution equation for the states (Eq. (2.82b)) can be formally integrated in analogy to Eq. (2.74) resulting in an equation of the form

$$\left|\Psi, A; (t)\right\rangle^F = \hat{U}^F(t, t_0)\left|\Psi, A; (t_0)\right\rangle^F \tag{2.86a}$$

$$\hat{U}^F(t, t_0) = \hat{\mathcal{T}} \exp\left[-\mathrm{i}\int_{t_0}^{t} dt \hat{H}^F_{\text{int}}\right]. \tag{2.86b}$$

A given initial state, formed at a time t_i and subsequently evolved by $\hat{U}^F$ can then be projected onto a complete basis of states, formed at a time t_f, after the scattering took place [21]

$$\sum_f |\Psi_f, A_f; (t_f)\rangle^{F\,F}\langle\Psi_f, A_f; (t_f)|\hat{U}^F(t_f, t_i)|\Psi, A_i; (t_i)\rangle^F$$
$$= \sum_f |\Psi_f, A_f; (t_f)\rangle^F \hat{U}^F_{fi}(t_f, t_i), \tag{2.87}$$

where we defined the evolution operator matrix elements

$$\hat{U}^F_{fi}(t_f, t_i) = {}^F\langle \Psi_f, A_f; (t_f)|\hat{U}^F(t_f, t_i)|\Psi, A; (t_i)\rangle^F. \tag{2.88}$$

The subscripts refer to the initial and final state, respectively. To capture an experimental scattering scenario as described above, one then has to consider states formed at $t_i \to -\infty$ and observed at $t_f \to \infty$. The scattering matrix in the Furry picture is thus recovered by the limit

$$\begin{aligned} S^F_{fi} &= U^F_{fi}(t_f \to \infty, t_i \to -\infty) \\ &= {}^F\langle \Psi_f, A_f; (t_f \to \infty)|\hat{S}^F|\Psi_i, A_i; (t_i \to -\infty)\rangle^F, \end{aligned} \tag{2.89}$$

in analogy to the ordinary QED result, obtained in the interaction picture [21]. The entries of the infinitely dimensional S-Matrix can then be understood as amplitudes of a given initial state to evolve into a specific final state. The scattering operator in the Furry picture is then in accordance with Eq. (2.74)

$$\hat{S}^F = \hat{\mathcal{T}} \exp\left[-\mathrm{i}\int_{-\infty}^{\infty} dt H^F_{\text{int}}(t)\right] = \hat{\mathcal{T}} \exp\left[-\mathrm{i}e\int d^4x\, \hat{\bar{\Psi}}^F \hat{\not{A}}^F_{\text{rad}}\, \hat{\Psi}^F\right], \tag{2.90}$$

The scattering operator in the Furry picture thus only describes the interaction with the quantized radiation field. From Eq. (2.90) one then constructs the usual perturbation series in orders of the exponential operator function. This perturbation series can then also be represented by Feynman graphs. However, all electron states in the Furry picture perturbation series are formed in the presence of the strong electromagnetic potential $A^\mu_{\text{ext}}(x)$. This particularity is conventionally indicated in the Feynman graph representation of the Furry picture, by drawing double lines for the electron states and propagators. This convention is adopted in the thesis and we will explicitly call attention to Feynman graphs which are drawn with single lines for the electron states, which are then ordinary perturbative QED graphs.

2.2.2.1 The Volkov Solution

According to the theory outlined in the previous section, we can take arbitrary external fields analytically into account in QED, as long as we can provide an exact solution of the Dirac equation (2.63a) in the given electromagnetic field. Unfortunately there exists only a fairly limited class of electromagnetic fields, the Dirac equation has been solved for analytically so far. One such known solution is given for the case of the external potential being a plane wave [21, 28]. For many applications, treating a laser field as a plane wave is a sufficiently good approximation. Furthermore, the plane wave solution is the leading order approximation to the electron state function, even for the external field being a focused Gaussian beam (see Appendix C). To obtain a solution of the Dirac equation one multiplies Eq. (2.63a) from the left with the operator $\left(\hat{\not{p}} - e\not{A}_L + m\right)$ to arrive at the second order differential equation

$$\left[\left(\hat{\not{p}} - e\not{A}_L\right)^2 - m^2 - \frac{i}{2}eF_{\mu\nu}\sigma^{\mu\nu}\right]\Psi(x) = 0, \tag{2.91}$$

where we used the antisymmetric tensor of the Dirac matrices defined in Appendix B. Eq. (2.91) is usually the starting point of the derivation. Recalling the Lorenz gauge condition we have $\partial_\mu A_L^\mu = 0$ and $\not{p}\not{p} = p^2$, the square operator term is evaluated to

$$\left(\hat{\not{p}} - e\not{A}_L\right)^2 = -\partial_\mu\partial^\mu - 2\mathrm{i}e\left(A_L^\mu\partial_\mu\right) + e^2A_L^2, \tag{2.92}$$

where the four dimensional unit matrix is not written explicitly. Due to the known plane wave solutions of the free Dirac equation (see Appendix B) it is sensible to expect the change in the wave function due to the external field to be summable in a prefactor, leading to the ansatz

$$\Psi_p(x) = \mathrm{e}^{-\mathrm{i}px}F_p(\eta). \tag{2.93}$$

Inserting this expression into Eq. (2.91) and expanding the square operator term as shown above one arrives at the equation

$$2\mathrm{i}(pk_L)F_p'(\eta) + \left[-2e\left(pA_L\right) + e^2A_L^2 - \mathrm{i}e\not{k}_L\left(\partial_\eta\not{A}(\eta)\right)\right]F_p(\eta) = 0. \tag{2.94}$$

This equation, however, is a simple first order differential equation for the prefactor $F_p(\eta)$ which is readily integrated to yield the expression

$$\Psi_p(x) = \mathrm{e}^{-\mathrm{i}px}\exp\left[-\mathrm{i}\int_0^\eta d\phi\left(\frac{e(pA_L(\phi))}{pk_L} - \frac{e^2A_L^2(\phi)}{2(pk_L)}\right) + e\frac{\not{k}_L\not{A}_L\,A(\eta)}{2(pk_L)}\right]\frac{u_p}{\sqrt{2\varepsilon V}}. \tag{2.95}$$

with a yet arbitrary spinor u_p and normalization factor $(2\varepsilon V)^{-1/2}$. The exponential series involving Dirac matrices is seen to vanish after its linear term, due to the relation $(\not{k}_L\not{A}_L)^2 = 0$. To eliminate all solutions of the second order Eq. (2.91), which are not solutions of the original first order Dirac Eq. (2.63a), we demand the solution to fulfill this first order equation at any point in space. For an arbitrarily small damping in the field $A_L(\eta)$ this request at $|\boldsymbol{r}| \to \infty$ goes over into

$$\left(\hat{\not{p}} - m\right)u_p = 0, \tag{2.96}$$

whence we conclude that the request of the constant spinor in Eq. (2.95) to be a solution of the free Dirac equation (see Appendix B) is already sufficient to ensure that the following wave functions are solutions of the Dirac equation in the presence of a plane wave

$$\Psi_p(x) = \left[1 + e\frac{\not{k}_L \not{A}_L(\eta)}{2(pk_L)}\right] \frac{u_p}{\sqrt{2\varepsilon V}} \mathrm{e}^{\mathrm{i}S_p(\eta)}. \tag{2.97}$$

This wave function is called *Volkov function* after D. Volkov who first published its derivation. The exponential phase given by

$$\begin{aligned} S_p(\eta) &= -px - g_p(\eta) \\ g_p(\eta) &= \int_0^\eta \mathrm{d}\phi \left(\frac{e(pA_L(\phi))}{pk_L} - \frac{e^2 A_L^2(\phi)}{2(pk_L)}\right). \end{aligned} \tag{2.98}$$

It is noteworthy that Eq. (2.98) is equivalent to the action of a classical electron in a plane wave fields given in Eq. (2.24) rendering the Volkov solution explicitly quasiclassical. It is further customary to assign a separate symbol to the combined matrix and exponential prefactors according to [29]

$$E_p(x) = \left[1 + e\frac{\not{k}_L \not{A}_L(\eta)}{2(pk_L)}\right] \mathrm{e}^{\mathrm{i}S_p(\eta)}, \tag{2.99}$$

which is called the *Ritus matrix*. To obtain the Dirac current associated with the Volkov functions we need the Dirac conjugate of Eq. (2.97)

$$\overline{\Psi}_p(x) = \frac{\bar{u}_p}{\sqrt{2\varepsilon V}} \overline{E}_p(x), \tag{2.100}$$

with the conjugate of the Ritus matrix defined in accordance with Eq. (2.99)

$$\overline{E}_p(x) = \left[1 + e\frac{\not{A}_L(\eta) \not{k}_L}{2(pk_L)}\right] \mathrm{e}^{-\mathrm{i}S_p(x)}. \tag{2.101}$$

In Eq. (2.102) we summarize the action of the Dirac operator on the Ritus matrices, the resulting explicit commutation relation with the contraction of the momentum with the Dirac matrices as well as the fact that the Ritus matrix and its Dirac conjugate are their respective inverse

$$(\mathrm{i}\not{\partial} - e\not{A}_L)\, E_p(x) = E_p(x)\not{p} \tag{2.102a}$$

$$\overline{E}_p(x)\, (\mathrm{i}\not{\partial} - e\not{A}_L) = -\not{p}\, \overline{E}_p(x) \tag{2.102b}$$

$$\left[\overline{E}(p,x), \not{p}\right] = \left[\not{p}, E(p,x)\right] = e\not{A}(x) - e\frac{A(x)p}{k_L p}\not{k}_L \tag{2.102c}$$

$$E_p(x)\overline{E}_p(x) = \mathbb{1}^4. \tag{2.102d}$$

We stress that the Ritus matrices depend nontrivially only on the laser phase η or equivalently the x^--coordinate. For later convenience we write the part of $E_p(x)$, depending solely on x^- explicitly

$$E_p(x^-) = \left[1 + e\frac{\not{k}_L \not{A}_L(x^-)}{2(pk_L)}\right] \mathrm{e}^{-\mathrm{i}(p^+x^- + g_p(x^-))}, \tag{2.103a}$$

$$\overline{E}_p(x^-) = \left[1 + e\frac{\not{A}_L(x^-)\not{k}_L}{2(pk_L)}\right] \mathrm{e}^{\mathrm{i}(p^+x^- + g_p(x^-))}, \tag{2.103b}$$

where we recall the discussion in connection to Eq. (2.5). The Volkov current is accordingly found to be

$$j_p^\mu(\eta) = \overline{\Psi}_p \gamma^\mu \Psi_p = \frac{1}{\varepsilon V}\left(p^\mu - eA_L^\mu(\eta) + k_L^\mu \left[e\frac{pA_L(\eta)}{pk_L} - \frac{e^2 A_L^2(\eta)}{2(pk_L)}\right]\right). \tag{2.104}$$

We note that the expression in round brackets is equivalent to the classical electron momentum in a plane wave from Eq. (2.31a). If we consider the external plane wave field to be confined in time such that $A_L(\eta \to \pm\infty) \to 0$, we infer that Volkov solutions, formed in the far past and future, are normalized to one particle per volume V. Hence, we can employ the usual state statistics for computing scattering probabilities for a pulsed field entering the Volkov solutions. Finally, owing to some discussion that was had on the topic, we state that the Volkov solutions are orthogonal and normalized according to [1, 29–31]

$$\frac{V}{(2\pi)^3} \int d\boldsymbol{x}\, \overline{\Psi}_{p'}(x)\gamma^0 \Psi_p(x) = \delta^{(3)}(\boldsymbol{p}' - \boldsymbol{p}). \tag{2.105}$$

It has recently been proven that the Volkov solutions fulfill the completeness relation [32]

$$\frac{V}{(2\pi)^3} \int d\boldsymbol{p}\, \overline{\Psi}_p(x)\gamma^0 \Psi_p(y) = \delta^{(3)}(\boldsymbol{x} - \boldsymbol{y}). \tag{2.106}$$

For the Ritus matrices Eqs. (2.99) and (2.101) this implies the orthogonality and completeness relations

$$\int \frac{d^4p}{(2\pi)^4} \overline{E}_p(x) E_p(y) = \delta^{(4)}(x-y), \int \frac{d^4x}{(2\pi)^4} \overline{E}_{p'}(x) E_p(x) = \delta^{(4)}(p'-p). \tag{2.107}$$

Having established the mentioned properties of the Volkov solutions we may feel free to use them as a basis for building a SF-QED field theory.

2.2.2.2 The Volkov Propagator

Irrespective of the picture in which the quantum dynamics are described, in addition to wave functions for the incoming and outgoing particles, in QED one is in need of the two-point Green's function or the propagator of the involved quantum fields. By virtue of the above given argument, that in tree level QED electromagnetic potentials do not interact, the dressed photon Green's function equals its free counterpart. The Green's function of a charged spinor field, on the other hand, is altered by including an external potential. In fact, the defining equation of the Green's function of the Dirac equation to an external plane wave potential $A_L(\eta)$ is [21]

$$(\mathrm{i}\not{\partial} - e\not{A} - m)\, G(x, y) = \delta^{(4)}(x - y). \tag{2.108}$$

The Green's function solving this equation can be expressed as [29, 33]

$$G(x, y) = \lim_{\epsilon\to 0} \int \frac{d^4p}{(2\pi)^4} E_p(x) \frac{\not{p} + m}{p^2 - m^2 + \mathrm{i}\epsilon} \overline{E}_p(y), \tag{2.109}$$

as can be checked by employing Eqs. (2.102a) and (2.102b). The pole prescription of Eq. (2.109) is the Feynman prescription of the free propagator [22]

$$G^{\text{free}}(x, y) = \lim_{\epsilon\to 0} \int \frac{d^4p}{(2\pi)^4} \frac{\not{p} + m}{p^2 - m^2 + \mathrm{i}\epsilon} \mathrm{e}^{-\mathrm{i}p(x-y)}, \tag{2.110}$$

to which the given dressed propagator reduces in the limit $A_L \to 0$. This prescription for bypassing the poles of Eq. (2.109) ensures that for $x^0 < y^0$, corresponding to propagation forward in time, only positive energy solutions ($\varepsilon > 0$) of Eq. (2.63a) enter Eq. (2.109), whereas in the opposite case $x^0 > y^0$, corresponding to propagation backwards in time, the negative energy components ($\varepsilon < 0$) are described.

Separating these two cases ab initio, we can split up the time ordered product $\hat{\mathcal{T}}\{\dots\}$ in the definition of the dressed propagator [21] into its respective forms

$$G(x, y) = -\mathrm{i}\langle 0|\hat{\mathcal{T}}\left\{\hat{\Psi}(y)\hat{\overline{\Psi}}(x)\right\}|0\rangle = \begin{cases} -\mathrm{i}\langle 0|\hat{\overline{\Psi}}(x)\hat{\Psi}(y)|0\rangle & \text{if } x_0 > y_0 \\ \mathrm{i}\langle 0|\hat{\Psi}(y)\hat{\overline{\Psi}}(x)|0\rangle & \text{if } x_0 < y_0 \end{cases}, \tag{2.111}$$

using the Dirac field operators Eqs. (2.73b) and (2.73c) expanded in the basis of Volkov functions. By virtue of the commutation relations Eq. (2.66a) one then finds a result equivalent to Eq. (2.109), as it has to be.

In a monochromatic external plane wave field the dressed electron propagator of SF-QED features infinitely many singularities depending on the number of photons absorbed from the external field [34, 35]. These resonances were addressed in numerous work employing the dressed electron propagator in monochromatic fields to compute numerous physical quantities such as lepton pair creation [36–39], lepton-lepton scattering [40, 41], resonant lepton-photon scattering (e.g. bremsstrahlung if the external photon is a nuclear Coulomb field photon) [42, 43] and, on a more fundamental level, those resonances were also investigated in the study of the electron self-energy [44]. These poles are located at the dressed mass m^*, which is discussed in the following section. In contrast to those previous works, there has been some effort to study SF-QED processes involving the dressed electron propagator in pulsed plane wave fields [45, 46]. There the authors, however, considered the strongly restrictive condition of temporally only mildly focussed ($\omega\tau \gg 1$) and low intense ($\xi \ll 1$) fields.

2.2.3 Interaction with a Monochromatic Laser Wave

In this chapter we outline the quantum analysis of the scattering of an electron from a monochromatic laser wave, in analogy to the discussion of Sect. 2.1.2. As discussed there, modeling the laser field as monochromatic allows for great simplifications in the calculations. In fact, most of the theoretical works on nonlinear Compton scattering, performed before this thesis was started, considered a monochromatic laser wave [29, 47–51]. There had been some work on electron scattering from a laser pulse of duration τ_L and frequency ω [19], but there the authors considered a pulse explicitly violating Eq. (1.9), i.e. a pulse containing many cycles of the carrier field. In this case, as one can see from Eq. (2.31a, 2.31b), the classical electron trajectory is strictly monotonic. Comparable to the Fourier decomposition of the radiation formula Eq. (2.17) it is possible to expand the Volkov states, and equally the dressed propagator in a Fourier series. To this end we consider a monochromatic laser wave of the form $\psi_A(\eta) = \sin(\eta)$ (the discussion for $\psi_A(\eta) = \cos(\eta)$ is analogous), whence the Volkov solutions becomes analytically integrable, yielding

$$\Psi_p(x) = \left[1 + e\frac{\not{k}_L \not{A}_L(\eta)}{2(pk_L)}\right] \frac{u_p}{\sqrt{2\varepsilon V}} \mathrm{e}^{-\mathrm{i}S_p^{\mathrm{m.c.}}(\eta)}, \tag{2.112}$$

with the monochromatic exponential $S_p^{\mathrm{m.c.}}(\eta) = \alpha\cos(\eta) + \beta\sin(2\eta) + qx$. In this expression we defined the quantities

$$q^\mu = p^\mu + \frac{m^2\xi^2}{4(pk_L)} k_L^\mu \tag{2.113a}$$

$$\alpha = -e\frac{pA_L}{pk_L} \tag{2.113b}$$

$$\beta = -\frac{e^2 A_L^2}{8pk_L}. \tag{2.113c}$$

The so-called dressed momentum q^μ contains an additional momentum component along the laser's wave vector k_L, which arises due to the non-vanishing average over the square contribution of the laser wave in the classical action and is thus attributed to the wiggling motion of the electron in the laser wave. The square of the dressed momentum yields the dressed mass m^* at which the poles of the electron propagator in a monochromatic laser wave are located

$$q^2 = m^2\left(1+\frac{m^2\xi^2}{2}\right) =: m^{*2}. \tag{2.114}$$

This dressed mass exceeds a free electron's rest mass by an intensity dependent term $m^{*2}-m^2 = m^2\xi^2/2$. This mass increase is caused by the periodic wiggling motion the electron undergoes inside the strong laser wave providing it with additional energy, which translates to an increased mass. We will find that when considering a laser pulse, the divergences from Eq. (2.109) are naturally regularized and the dressed mass loses its unambiguous meaning. Though this observation does not render the concept of a dressed mass obsolete, it is clear that it will definitely need further investigation (see Sect. 3.5 and [52]). The expression for the Volkov states (2.112) allows for the deployment of the generating function of Bessel functions analogous to Eq. (2.40) from classical electrodynamics. Utilizing the relation

$$\sin^i(\eta)\mathrm{e}^{-\mathrm{i}(\alpha\sin(\eta)+\beta\sin(2\eta))} = \sum_{n=-\infty}^{\infty} C_{i,n}\mathrm{e}^{-\mathrm{i}n\eta} \tag{2.115}$$

where the coefficients are defined according to [29]

$$C_{i,n} = \frac{1}{2\pi}\int d\eta' \sin(\eta')\mathrm{e}^{\mathrm{i}(\alpha\sin(\eta')+\beta\sin(2\eta')-n\eta')} \tag{2.116}$$

allows then for a Fourier expansion of Eq. (2.112) resulting in

$$\Psi_p(x) = \sum_{n=-\infty}^{\infty}\left[C_{0,n} + e\frac{\not{k}_L\not{A}_L}{2(pk_L)}C_{1,n}\right]\frac{u_p}{\sqrt{2q^0V}}\mathrm{e}^{-\mathrm{i}(q+nk_L)x}. \tag{2.117}$$

The replacement $\varepsilon \to q^0$ in the normalization of the monochromatic wave function is due to the special form of an infinitely stretched external laser field. The wave functions are required to be normalized to one particle per normalization volume and as in the monochromatic case one cannot perform the limit $A_L^\mu(\eta \to \pm\infty) = 0$, one has to average this quantity over one laser cycle, yielding

$$\bar{j}_p^\mu(\eta) = \frac{1}{q^0V}\left(p^\mu - k_L^\mu\frac{m^2\xi^2}{4(pk_L)}\right) = \frac{q^\mu}{q^0V}, \tag{2.118}$$

which again corresponds to one particle per volume V. Equation (2.117) is now well suited for computing scattering matrix elements, since the exponential factors will always cancel the four dimensional space-time integration, to give a momentum conserving δ-function of the form $\delta^{(4)}(\sum_{\text{in}} q - \sum_{\text{out}} q)$, i.e., where in the sums for spinor particles there have to be inserted the dressed momenta q, whereas uncharged particles, such as photons, enter with their ordinary free momentum. Illustrating this concept at the exemplary process of an electron of initial momentum p_i being scattered by a monochromatic laser wave, described by $A_L(\eta)$ into a final momentum state q_f upon emission of a single photon with wave vector k_1, the scattering matrix amplitude is given by [29]

$$S_{fi} = -\mathrm{i}\frac{\sqrt{2\pi}e}{\sqrt{\omega_1 V}} \int \mathrm{d}^4x\, \overline{\Psi}_{p_f}(x) \not{\epsilon}_1^* \mathrm{e}^{\mathrm{i}k_1 x} \Psi_{p_i}(x), \tag{2.119}$$

where for the emitted photon the free wave function

$$A_1^\mu = \frac{\sqrt{4\pi}}{\sqrt{2\omega_1 V}} \epsilon_1^\mu \mathrm{e}^{\mathrm{i}k_1 x} \tag{2.120}$$

solving Eq. (2.1) is employed with the polarization index not written explicitly. Expanding this expression now into a Fourier series analogously to Eq. (2.117) one obtains the expression

$$S_{fi} = -\mathrm{i}\frac{\sqrt{\pi}e(2\pi)^4}{\sqrt{2\omega_1 q_i^0 q_f^0 V^3}} \sum_{n=-\infty}^{\infty} \bar{u}_{p_f} M_n u_{p_i} \delta^{(4)}(q_i + nk_L - k_1 - q_f), \tag{2.121}$$

with the reduced matrix elements given by

$$M_n = \left[\not{\epsilon}_1^* C_{0,n} + e\left(\frac{\not{A}_L \not{k}_L}{2(p_f k_L)} + \frac{\not{k}_L \not{A}_L}{2(p_i k_L)} \right) C_{1,n} - \frac{e^2 A_L^2 (k_L \epsilon_1^*)}{2(p_i k_L)(p_f k_L)} \not{k}_L C_{2,n} \right]. \tag{2.122}$$

The scattering matrix element is then easily translated into an emission probability per unit time by taking its modulus square, summing and averaging over all outgoing and incoming particles' spins and polarizations, respectively, and multiplying the result by the phase space of the final state's particles. The result of this procedure is given by

$$\begin{aligned} \frac{1}{T} dW^{\text{m.c.}} &= \frac{1}{2} \sum_{\{\sigma,\lambda\}} \left| S_{fi} \right|^2 \frac{d\boldsymbol{k}_1 V}{(2\pi)^3} \frac{d\boldsymbol{p}_f V}{(2\pi)^3} \\ &= \frac{e^2}{16\pi\omega_1 q_i^0 q_f^0} \sum_{\{\sigma,\lambda\}} \sum_{n=-\infty}^{\infty} \left| \bar{u}_{p_f} M_n u_{p_i} \right|^2 d\boldsymbol{k}_1 d\boldsymbol{p}_f \delta^{(4)}(q_i + nk_L - k_1 - q_f), \end{aligned} \tag{2.123}$$

where the square of the four dimensional δ-function yielded the customary factor $(VT)/(2\pi)^4$, $\sum_{\{\sigma,\lambda\}}$ means the summation over all incoming and outgoing polarization and spin quantum numbers and the additional factor $1/2$ turns the sum over the initial state's electron spins into an average. In Eq. (2.123) the three spatial δ-functions fix the final electron's spatial momentum, fixing final electron's energy to the value $q_f^{0\,2} = q_i^{0\,2} + n\omega_L^2 + \omega_1^2 + 2\left(\boldsymbol{q}_i\boldsymbol{k}_L - \boldsymbol{q}_i\boldsymbol{k}_1 - n\boldsymbol{k}_L\boldsymbol{k}_1\right)$. This procedure introduces a conversion factor $\left|\partial \boldsymbol{q}_f/\partial \boldsymbol{p}_f\right|^{-1} = \left|\partial q_f^{\|}/\partial p_f^{\|}\right|^{-1}$ into Eq. (2.123). The fourth δ-function is customarily used to fix the outgoing photon's energy to the harmonic frequencies

$$\omega_{1,n}^{\text{m.c.}} = \frac{n(p_i k_L)}{(q_i + nk_L)n_1}, \tag{2.124}$$

with an integer $n \in [0, \infty]$. In the classical limit $k_1 k_L \ll p_i k_L$ this expression goes over to the classical harmonic formula (2.43), whence one concludes that the ordinary Doppler shift can be recovered only in the limit $\xi \ll 1$. Using the fourth δ-function in Eq. (2.123) to fix ω_1 to the values of Eq. (2.124), introduces an additional factor of the form [29, 53]

$$\left|\frac{d(q_f^0 + \omega_1)}{d\omega_1}\right| = \frac{(q_i + nk_L)n_1}{q_f^0}. \tag{2.125}$$

As a result of the described steps we obtain the final expression for the single photon emission probability of an electron scattered from a monochromatic laser field per unit time and solid angle

$$\frac{1}{T}\frac{dW^{\text{m.c.}}}{d\Omega_1} = \frac{e^2\omega_1}{16\pi q_i^0 q_f^0 \left|\frac{d(q_f^0+\omega_1)}{d\omega_1}\right|} \sum_{\{\sigma,\lambda\}} \sum_{n=-\infty}^{\infty} \left|\bar{u}_{p_f} M_n u_{p_i}\right|^2. \tag{2.126}$$

In the same fashion one can easily obtain expressions for higher order SF-QED processes in a monochromatic laser wave.

2.2.4 Interaction with a Laser Pulse

Concerning temporal compression, there has been an increasingly fast growing number of works analyzing QED amplitudes of electrons interacting with plane wave laser fields of arbitrary strength. These works, which are applicable to the realm of pulse durations distinguished by Eq. (1.9), were performed for single photon emission [20, 51, 54–58] and recently also for two photon emission [59, 60]. This family of calculations, applicable to the rapidly evolving regime of few-cycle laser pulses, is a research field of swiftly increasing interest and importance.

It is this research field the present thesis is dedicated to.

References

1. L.D. Landau, E.M. Lifschitz, *Lehrbuch der Theoretischen Physik - Band 2: Klassische Feldtheorie* (Akademie Verlag, Berlin, 1997)
2. T. Brabec, F. Krausz, Nonlinear optical pulse propagation in the single-cycle regime. Phys. Rev. Lett. **78**, 3282 (1997)
3. T. Heinzl, A. Ilderton, M. Marklund, Finite size effects in stimulated laser pair production. Phys. Lett. B **692**, 250 (2010)
4. A. Di Piazza, K.Z. Hatsagortsyan, C.H. Keitel, Strong signatures of radiation reaction below the radiation-dominated regime. Phys. Rev. Lett. **102**, 254802 (2009)
5. A. Di Piazza, K.Z. Hatsagortsyan, C.H. Keitel, Quantum radiation reaction effects in multiphoton compton scattering. Phys. Rev. Lett. **105**, 220403 (2010)
6. I.V. Sokolov, J.A. Nees, V.P. Yanovsky, N.M. Naumova, G. Mourou, Emission and its back-reaction accompanying electron motion in relativistically strong and QED-strong pulsed laser fields. Phys. Rev. E **81**, 036412 (2010)
7. M. Tamburini, F. Pegoraro, A. Di Piazza, C.H. Keitel, T.V. Liseykina, A. Macchi, Radiation reaction effects on electron nonlinear dynamics and ion acceleration in laser solid interaction. Nucl. Instrum. Meth. A **653**, 181 (2011)
8. D.J. Jackson, *Classical Electrodynamics* (Wiley, New York, 1975)
9. V.N. Baier, V.M. Katkov, V.M. Strakhovenko, *Electromagnetic Processes at High Energies in Oriented Single Crystals* (World Scientific, Singapore, 1994)
10. J.H. Eberly, A. Sleeper, Trajectory and mass shift of a classical electron in a radiation pulse. Phys. Rev. **176**, 1570 (1968)
11. E.S. Sarachik, G.T. Schappert, Classical theory of the scattering of intense laser radiation by free electrons. Phys. Rev. D **1**, 2738 (1970)
12. Y.I. Salamin, F.H.M. Faisal, Harmonic generation by superintense light scattering from relativistic electrons. Phys. Rev. A **54**, 4383 (1996)
13. J.W. Meyer, Covariant classical motion of electron in a laser beam. Phys. Rev. D **3**, 621 (1971)
14. F.V. Hartemann, D.J. Gibson, A.K. Kerman, Classical theory of Compton scattering: assessing the validity of the Dirac-Lorentz equation. Phys. Rev. E **72**, 026502 (2005)
15. Y.I. Salamin, F.H.M. Faisal, Generation of Compton harmonics by scattering linearly polarized light of arbitrary intensity from free electrons of arbitrary initial velocity. J. Phys. A - Math. Gen. **31**, 1319 (1998)
16. H.R. Reiss, Absorption of light by light. J. Math. Phys. **3**, 59 (1962)
17. Y.I. Salamin, G.R. Mocken, C.H. Keitel, Electron scattering and acceleration by a tightly focused laser beam. Phys. Rev. ST - Accel. Beams **5**, 101301 (2002)
18. L.W. Davis, Theory of electromagnetic beams. Phys. Rev. A **19**, 1177 (1979)
19. N.B. Narozhnyi, M.S. Fofanov, Photon emission by electron during collision with short focused laser pulses. J. Exp. Theo. Phys. **110**, 26 (1996)
20. D. Seipt, B. Kämpfer, Nonlinear Compton scattering of ultrashort intense laser pulses. Phys. Rev. A **83**, 022101 (2011)
21. L.D. Landau, E.M. Lifschitz, *Lehrbuch der Theoretischen Physik - Band 4: Quantenelektrodynamik* (Akademie Verlag, Berlin, 1991)
22. M.E. Peskin, D.V. Schroeder, *An introduction to Quantum Field Theory* (Westview Press, Boulder, 1995)
23. W. Greiner, J. Reinhardt, *Quantum Electrodynamics* (Springer, Heidelberg, 2002)
24. E.S. Fradkin, D.B. Gitman, S.M. Shvartsman, *Quantum Electrodynamics with Unstable Vacuum* (Springer, Heidelberg, 1991)
25. F. Mandl, G. Shaw, *Quantum field theory* (Wiley, New York, 1984)
26. P.A.M. Dirac, The quantum theory of the electron. P. Roy. Soc. Lond. A Mat. **117**, 610 (1928)
27. W.H. Furry, On bound states and scattering in positron theory. Phys. Rev. **81**, 115 (1951)
28. D.M. Volkov, Class of solutions of Dirac's equation. Z. Phys. **94**, 250 (1935)
29. V.I. Ritus, Quantum effects of the interaction of elementary particles with an intense electromagnetic field. J. Russ. Laser Res. **6**, 497 (1985)

30. H. Prakash, N. Chandra, Use of Volkov solutions of electron wave equations in electromagnetic wave for scattering problems. Nuovo Cimento B **55**, 404 (1968)
31. S. Zakowicz, Square-integrable wave packets from the Volkov solutions. J. Math. Phys. **46**, 032304 (2005)
32. M. Boca, On the properties of the Volkov solutions of the Klein Gordon equation. J. Phys. A - Math. Gen. **44**, 445303 (2011)
33. H.R. Reiss, J.H. Eberly, Green's function in intense-field electrodynamics. Phys. Rev. **151**, 1058 (1966)
34. V.P. Oleinik, Resonance effects in field of an intense laser beam. Sov. Phys. JETP **25**, 697 (1967)
35. W. Becker, H. Mitter, Modification of the quasi-levels of an electron in a laser field due to radiative corrections. J. Phys. A - Math. Gen. **9**, 2171 (1976)
36. C. Müller, K.Z. Hatsagortsyan, C.H. Keitel, Muon pair creation from positronium in a circularly polarized laser field. Phys. Rev. D **74**, 074017 (2006)
37. C. Müller, K.Z. Hatsagortsyan, C.H. Keitel, Muon pair creation from positronium in a linearly polarized laser field. Phys. Rev. A **78**, 033408 (2008)
38. A. Di Piazza, E. Lötstedt, A.I. Milstein, C.H. Keitel, Barrier control in tunneling e^+-e^- photoproduction. Phys. Rev. Lett. **103**, 170403 (2009)
39. K. Krajewska, Electron-positron pair creation and Oleinik resonances. Laser Phys. **21**, 1275 (2011)
40. P. Panek, J.Z. Kamiński, F. Ehlotzky, Analysis of resonances in Møller scattering in a laser field of relativistic radiation power. Phys. Rev. A **69**, 013404 (2004)
41. P. Panek, J.Z. Kaminski, F. Ehlotzky, Moller scattering in a laser field of relativistic radiation power. Laser Phys. **14**, 1200 (2004)
42. S. Schnez, E. Lötstedt, U.D. Jentschura, C.H. Keitel, Laser-assisted bremsstrahlung for circular and linear polarization. Phys. Rev. A **75**, 053412 (2007)
43. E. Lötstedt, U.D. Jentschura, C.H. Keitel, Evaluation of laser-assisted Bremsstrahlung with Dirac-Volkov propagators. Phys. Rev. Lett. **98**, 043002 (2007)
44. J.H. Eberly, H.R. Reiss, Electron self-energy in intense plane-wave field. Phys. Rev. **145**, 1035 (1966)
45. E. Padusenko, S. Roshchupkin, Resonant scattering of a lepton by a lepton in the pulsed light field. Laser Phys. **20**, 2080 (2010)
46. A. Voroshilo, S. Roshchupkin, V. Nedoreshta, Resonant scattering of photon by electron in the presence of the pulsed laser field. Laser Phys. **21**, 1675 (2011)
47. L.S. Brown, T.W.B. Kibble, Interaction of intense laser beams with electrons. Phys. Rev. **133**, A705 (1964)
48. A.I. Nikishov, V.I. Ritus, Quantum processes in the field of a plane electromagnetic wave and in a constant field.1. Sov. Phys. JETP **19**, 529 (1964)
49. V.N. Baier, V.M. Katkov, A.I. Milshtein, V.M. Strakhovenko, Contribution to theory of quantum processes in field, of an intense electromagnetic wave. J. Exp. Theo. Phys. **69**, 783 (1975)
50. D.Y. Ivanov, G.L. Kotkin, V.G. Serbo, Complete description of polarization effects in emission of a photon by an electron in the field of a strong laser wave. Eur. Phys. J. C **36**, 127 (2004)
51. C. Harvey, T. Heinzl, A. Ilderton, Signatures of high-intensity Compton scattering. Phys. Rev. A **79**, 063407 (2009)
52. C. Harvey, T. Heinzl, A. Ilderton, M. Marklund, The intensity dependent mass shift: existence, universality and detection. Phys. Rev. Lett. **109**, 100402 (2012)
53. J.M. Jauch, F. Rohrlich, *Theory of Photons and Electrons* (Springer, Heidelberg, 1976)
54. M. Boca, V. Florescu, Nonlinear Compton scattering with a laser pulse. Phys. Rev. A **80**, 053403 (2009)
55. T. Heinzl, D. Seipt, B. Kämpfer, Beam-shape effects in nonlinear Compton and Thomson scattering. Phys. Rev. A **81**, 022125 (2010)
56. F. Mackenroth, A. Di Piazza, Nonlinear Compton scattering in ultrashort laser pulses. Phys. Rev. A **83**, 032106 (2012)

57. K. Krajewska, J.Z. Kamiński, Compton process in intense short laser pulses. Phys. Rev. A **85**, 062102 (2012)
58. F. Mackenroth, A. Di Piazza, Reply to "Comment on 'Nonlinear Compton scattering in ultrashort laser pulses'". Phys. Rev. A **85**, 046102 (2012)
59. D. Seipt, B. Kämpfer, Two-photon Compton process in pulsed intense laser fields. Phys. Rev. D **85**, 101701 (2012)
60. F. Mackenroth, A. Di Piazza, Nonlinear double Compton scattering in the full quantum regime. Phys. Rev. Lett. **110**, 070402 (2013)

Chapter 3
Nonlinear Single Compton Scattering

Nature uses only the longest threads to weave her patterns, so that each small piece of her fabric reveals the organization of the entire tapestry.

Richard P. Feynman

3.1 Introduction

The lowest order of the SF-QED perturbation series is described by Furry picture Feynman diagrams with only one vertex. The only possible scattering process with only an electron in the initial state is an electron with initial momentum $p_i^\mu = (\varepsilon_i, \boldsymbol{p}_i)$ emitting a single photon with wave vector $k_1^\mu = \omega_1 n_1$ and thereby changing its momentum to $p_f^\mu = (\varepsilon_f, \boldsymbol{p}_f)$. The corresponding Feynman diagram is shown in Fig. 3.1. Other possible lowest order SF-QED processes such as pair creation by a high energetic photon or the corresponding annihilation process, are connected to this process by crossing symmetry [1]. In field free QED a single vertex Feynman diagram does not correspond to a physical process, since energy momentum conservation requires

$$p_i = p_f + k_1. \tag{3.1}$$

Squaring this relation gives the condition $(p_i k_1) = \varepsilon_i \omega_1 (1 - \boldsymbol{\beta}_i \boldsymbol{n}_1) = 0$, which can only be satisfied for $\omega_1 = 0$. A free electron thus cannot emit or absorb a free photon.

In SF-QED, however, the situation is fundamentally different, since the electron states take an external laser field into account to all orders, which provides additional momentum and allows the emission of a single photon even at tree-level as denoted by the double line in Fig. 3.1 for the electron state. In Chap. 1 it was mentioned that this process has been abundantly investigated in classical electrodynamics (called *nonlinear Thomson scattering*) as well as in the quantum regime (NSCS). Albeit, the

K. F. Mackenroth, *Quantum Radiation in Ultra-Intense Laser Pulses*, Springer Theses, DOI: 10.1007/978-3-319-07740-6_3,

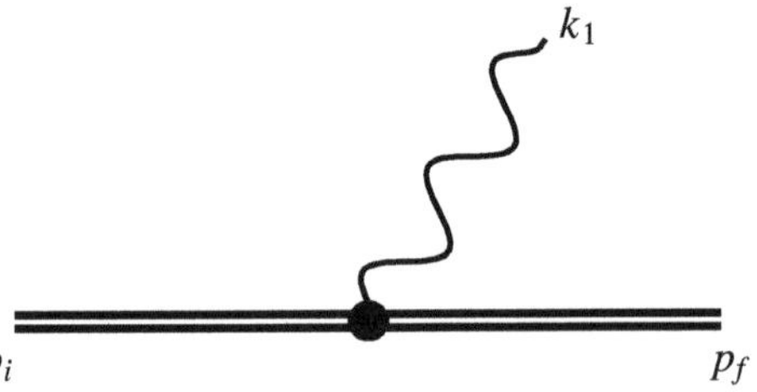

Fig. 3.1 Furry picture Feynman diagram of NSCS

works cited there were mostly performed under the assumption of a monochromatic external laser wave. In this chapter we will investigate NSCS in the presence of a laser pulse with arbitrary temporal shaping. The interest in this type of process has been developing quite recently (see Sect. 2.2.4). We additionally note that, as long as we restrict ourselves to values of the quantum parameter not exceeding $\chi \sim 1$, the main contribution to the overall quantum radiation stems from NSCS. Two-photon emission is roughly $\alpha_{QED} = e^2/\hbar c \approx 1/137$ times smaller than the process considered here [2].

3.2 Matrix Element and Transition Probability

The matrix element of the process depicted in Fig. 3.1 is given by

$$S_{fi} = -\mathrm{i}\frac{\sqrt{4\pi}e}{\sqrt{8\omega_1\varepsilon_i\varepsilon_f V^3}} \int \mathrm{d}^4x\ \overline{\Psi}_{p_f}(x)\not{\epsilon}_1^* \mathrm{e}^{\mathrm{i}k_1x}\Psi_{p_i}(x), \tag{3.2}$$

where $\Psi_{p_i}(x)$ ($\overline{\Psi}_{p_f}(x)$) is the Volkov wave function from Eq. (2.98) to the initial momentum p_i^μ (its conjugate to the final momentum p_f^μ). Due to the fact that the field is modeled by a plane wave, in Eq. (3.2) all but one coordinate dependencies may be integrated out to give

$$\begin{aligned} S_{fi} &= \mathcal{N}_{\mathrm{NSCS}}\ \bar{u}_{p_f}\ M_{fi}\ u_p\ \delta^{(3)}\left(p_i^\perp - k_1^\perp - p_f^\perp\right), \\ \mathcal{N}_{\mathrm{NSCS}} &= -\mathrm{i}\frac{e\sqrt{4\pi}(2\pi)^3}{\sqrt{8\omega_1\ \varepsilon_f\ \varepsilon_i\ V^3}}. \end{aligned} \tag{3.3}$$

Here the notion $\delta^{(3)}\left(p_i^\perp - k_1^\perp - p_f^\perp\right)$ refers to the energy momentum conservation laws for the three momentum components, perpendicular to p^+ (see Sect. 1.2). The reduced matrix element in Eq. (3.3) is given by

$$M_{fi} = \sum_{i=0}^{2} \tilde{\Gamma}_i f_i. \tag{3.4}$$

Here we defined the dynamic integrals

$$f_i = \int_{-\infty}^{\infty} dx^- \, \psi_{\mathcal{A}}^i(x^-) \mathrm{e}^{-\mathrm{i}S_{p_i,p_f}^{k_1}(x^-)} \tag{3.5}$$

$$S_{p_i,p_f}^{k_1}(x^-) = g_{p_i,p_f}^{k_1}(x^-) + k_L^+ \gamma_V \tag{3.6}$$

$$g_{p_i,p_f}^{k_1}(x^-) = k_L^+ \int_{-\infty}^{x^-} dc^- \alpha_V \psi_{\mathcal{A}}(c^-) + \beta_V \psi_{\mathcal{A}}^2(c^-), \tag{3.7}$$

with the exponential phase $S_{p_i,p_f}^{k_1}(x^-)$ describing an electron changing its momentum from p_i^μ to p_f^μ upon the emission of a photon with wave vector k_1^μ. The definition (3.6) contains the three parameters

$$\alpha_V = e\left(\frac{p_i A}{p_i k_L} - \frac{p_f A}{k_L p_f}\right) \tag{3.8a}$$

$$\beta_V = \frac{e^2 A_L^2}{2}\left(\frac{k_L k_1}{(k_L p_f)(p_i k_L)}\right) \tag{3.8b}$$

$$\gamma_V = -\frac{k_1 p_i}{k_L p_f}. \tag{3.8c}$$

The matrix prefactors in Eq. (3.4) are given by the expressions

$$\tilde{\Gamma}_0 = \not{\epsilon}_1^* \tag{3.9a}$$

$$\tilde{\Gamma}_1 = e\left(\frac{\not{A}_L \not{k}_L \not{\epsilon}_1^*}{2(k_L p_f)} + \frac{\not{\epsilon}_1^* \not{k}_L \not{A}_L}{2(p_i k_L)}\right) \tag{3.9b}$$

$$\tilde{\Gamma}_2 = -\frac{e^2 A_L^2 \, (k\epsilon_1^*)}{2(p_i k_L)(k_L p_f)} \not{k}_L\,, \tag{3.9c}$$

where $\not{A}_L = A_L \not{\epsilon}_L$. We can trade the last parameter Eq. (3.8c) for a new integration and a δ-function in (3.5) to find

$$f_i = k_L^+ \int_{-\infty}^{\infty} dx^- ds \, \psi_{\mathcal{A}}^i(x^-) \mathrm{e}^{-\mathrm{i}(g_{p_i,p_f}^{k_1}(x^-) + k_L^+ s)} \delta(p_i^+ - s k_L^+ - k_1^+ - p_f^+). \tag{3.10}$$

The advantage of this representation of the dynamic integrals is that it introduces a four dimensional energy momentum conservation into the scattering matrix element

$$S_{fi} \propto \int ds \ldots \delta^{(4)}(p_i - sk_L - k_1 - p_f). \tag{3.11}$$

This form of the the matrix element is also widely used [3–6] and can equally be obtained by means of Fourier transformation. Equation (3.11) allows for a physical interpretation of the parameter γ_V, as a continuous generalization of the number of laser mode photons absorbed by the electron. In the present work, however, we are going to base our analyses on Eq. (3.3). Comparing Eq. (3.6) to (2.24) we find that the exponential phase of the dynamic integrals is given by the difference of classical action functionals for an electron in the presence of a plane wave laser field [7, 8] and a photon emission term

$$S^{k_1}_{p_i,p_f}(x^-) = S^{\text{class}}_{p_i}(x^-) - S^{\text{class}}_{p_f}(x^-) - k_1 x. \tag{3.12}$$

In this sense the scattering matrix element (3.3) is often referred to as quasi-classical. The dynamic integral f_0 is divergent because it does not contain the shape function $\psi_\mathcal{A}(x^-)$ as a preexponential, which goes to zero in the considered case of a pulsed laser field. The integrand thus does not tend to zero for $x^- \to \pm\infty$. It can, however, be decomposed into a linear combination of the other two dynamic integrals, which are well defined due to the preexponential function $\psi_\mathcal{A}(x^-)$. By partial integration we find

$$\begin{aligned}\int_{-\infty}^{\infty} dx^- \mathrm{e}^{-\mathrm{i}g^{k_1}_{p_i,p_f}(x^-) - \mathrm{i}\gamma_V k_L^+ x^-} &= -\int_{-\infty}^{\infty} dx^- \frac{\partial_{x^-} g^{k_1}_{p_i,p_f}(x^-)}{\gamma_V k_L^+} \mathrm{e}^{-\mathrm{i}g^{k_1}_{p_i,p_f}(x^-) - \mathrm{i}\gamma_V k_L^+ x^-} \\ &= f_0 = -(\alpha_V f_1 + \beta_V f_2)\,.\end{aligned} \tag{3.13}$$

This decomposition can also be understood as a manifestation of gauge invariance [5], which requires

$$S_{fi}(\epsilon_1^\mu \to k_1^\mu) = 0. \tag{3.14}$$

To follow this argument one has to take advantage of the properties of the Ritus matrices stated in Eq. (2.103), the energy momentum conservation of Eq. (3.11) and the algebraic relations (B.10). Using these ingredients in Eq. (3.14) one finds

$$\begin{aligned}S_{fi}(\epsilon_1^\mu \to k_1^\mu) &\propto \int dx^-\, \overline{\Psi}_{p_f}(x^-) \not{k}_1 \Psi_{p_i}(x^-) \mathrm{e}^{-\mathrm{i}S^{k_1}_{p_i,p_f}(x^-)} \\ &= \int dx^-\, \bar{u}_{p_f} \overline{E}(p_f, x)(\not{p}_i - \gamma_V \not{k}_L - \not{p}_f) E(p_i, x) u_{p_i} \mathrm{e}^{-\mathrm{i}S^{k_1}_{p_i,p_f}(x^-)} \\ &= -\int dx^-\, \bar{u}_{p_f} \left[\alpha_V \psi_\mathcal{A}(x^-) + \beta_V \psi_\mathcal{A}(x^-) + \gamma_V\right] \not{k}_L u_{p_i} \mathrm{e}^{-\mathrm{i}S^{k_1}_{p_i,p_f}(x^-)}\end{aligned} \tag{3.15}$$

From the last line of the previous equation we arrive at a condition equivalent to Eq. (3.13).

The energy momentum conservation of Eq. (3.2) differs from the result for a monochromatic external field in that it is only three dimensional, whereas the result for a monochromatic field of Sect. 2.2.3 contains a four dimensional energy momentum conserving δ-function. This difference can be interpreted as the absence of a universal dressed mass as a contrast to the monochromatic case (see Sect. 3.5). We point out that three energy momentum conservation laws suffice to fully determine the electron's four momentum after the scattering, since this has to additionally fulfill the mass shell condition $p_f^2 = m^2$. By virtue of this condition one finds, that the x^--component of this final momentum needs to satisfy $p_f^- > 0$, whence one can determine the maximum energy which can be emitted into a direction $\boldsymbol{n}_1$

$$\omega_1^{\mathrm{Max}} = \frac{\varepsilon_i(1+\beta_i)}{1-\boldsymbol{n}_L\boldsymbol{n}_1}. \tag{3.16}$$

From the four dimensional energy momentum conservation of Eq. (3.11) we read off the final electron energy

$$\varepsilon_f = \varepsilon_i - \omega_1 + \frac{k_1 p_i}{n_L(p_i - k_1)}. \tag{3.17}$$

To include the regularization of f_0 according to Eq. (3.13) in the scattering matrix element Eq. (3.3) we only have to implement one redefinition. Instead of Eq. (3.4) the reduced matrix element has to be expressed as

$$M_{fi} = \sum_{i=1}^{2} \Gamma_i f_i, \tag{3.18}$$

with the dynamic integrals of Eq. (3.5) and the coefficients containing the Dirac structure of the scattering matrix element

$$\Gamma_1 = \frac{\not{A}_L \not{k}_L \not{\epsilon}_1^*}{2(k_L p_f)} + \frac{\not{\epsilon}_1^* \not{k}_L \not{A}_L}{2(p_i k_L)} - \frac{\alpha_V}{\gamma_V}\not{\epsilon}_1^* \tag{3.19}$$

$$\Gamma_2 = \frac{e^2 A^2 \left(k_L \epsilon_1^*\right)}{2(p_i k_L)(k_L p_f)}\not{k}_L - \frac{\beta_V}{\gamma_V}\not{\epsilon}_1^*. \tag{3.20}$$

Using this expression and taking the square of the δ-function following the arguments given in Appendix A, we can express the differential emission probability as

$$\frac{dW}{d\omega_1 d\Omega_1} = \frac{d^3\boldsymbol{p}_f V}{(2\pi)^3}\frac{\omega_1^2 V}{(2\pi)^3}\frac{1}{2}\sum_{\{\sigma,\lambda\}}\left|S_{fi}\right|^2 = \frac{e^2\,\omega_1}{(4\pi)^2 p_f^- p_i^-}\frac{1}{2}\sum_{\{\sigma,\lambda\}}\left|\bar{u}_{p_f} M_{fi} u_{p_i}\right|^2. \tag{3.21}$$

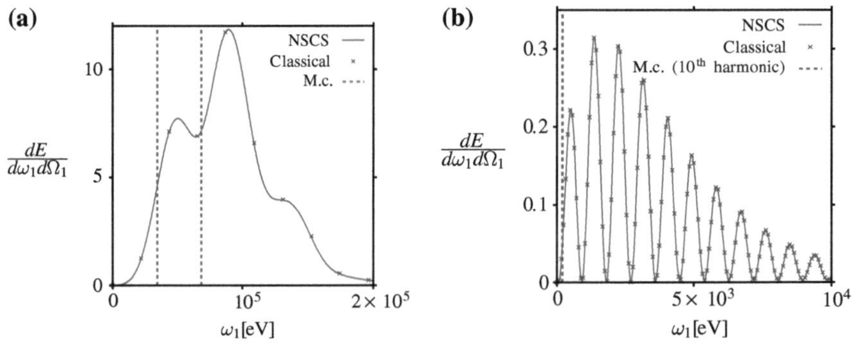

Fig. 3.2 Comparison of energy emission spectra, obtained via numerical integration of Eq. (3.21), to classical and monochromatic results for the interaction of an electron with initial energy ε_i colliding head on with a laser pulse described by Eq. (2.6) with $n_C = 2$. It is set $\varepsilon_i = 50$ MeV and $I_L = 2 \times 10^{18}$ W/cm² ($\xi = 1, \chi \approx 5 \times 10^{-4}$, see Fig. 3.2a) and $\varepsilon_i = 5$ MeV and $I_L = 10^{20}$ W/cm² ($\xi = 7, \chi \approx 4 \times 10^{-4}$, see Fig. 3.2b). The spectra are given in units of [sr^{-1}]

In this expression $\sum_{\{\sigma,\lambda\}}$ as usual means the summation over all discrete incoming and outgoing polarization and spin quantum numbers, where the additional factor 1/2 in front of the sum turns the summation over the incoming spin quantum numbers into an average. The explicit result of this procedure was given in several recent publications [6, 9, 10]. The three dimensional energy momentum conservation resulting from $|S_{fi}|^2$, according to Eq. (A.11), is here used to fix the electron's final momentum p_f^μ. Equation (3.21) provides the probability of an electron, entering a plane wave laser field $A_L^\mu(x^-)$ with an initial momentum p_i^μ to emit a photon with wave vector k_1^μ. The dynamics of this process are fully encoded in the dynamic integrals f_i, given in (3.5). In general, these integrals are not analytically integrable. Consequently, to obtain an emission probability from Eq. (3.21) we will evaluate them by a numeric quadrature scheme, described in Chap. 5. However, their properties can be investigated in some limiting cases. In Appendix D we sketch how the asymptotic emission probabilities of Eq. (3.21) are recovered in the perturbative (Sect. D.1), the classical (Sect. D.3) and the monochromatic limit (Sect. D.2). The main results of these somewhat lengthy and technical discussions, however, can be neatly summarized in a graphical display. As numerical results in the following we present, in accordance with the discussion of Appendix A, energy emission spectra $dE/d\omega_1 d\Omega_1$, observed in a reference frame according to Fig. 2.1. As we also consider only linearly polarized laser pulses, the direction in which we observe the radiation emitted is uniquely distinguished by the angle ϑ_1 between $\boldsymbol{n}_1$ and $\boldsymbol{n}_L$ (see discussion subsequent to Fig. 2.2). In connection with Eq. (1.7) we found that the classical limit is recovered in the regime $\chi \ll 1$. In this regime the emission spectra, obtained from Eq. (3.21) must agree with a classical result obtained from Eq. (2.35). The formal equivalence of these two expression in the classical regime is shown in Sect. D.3. To demonstrate it numerically, in Fig. 3.2 we show the comparison of the two results, both obtained by numerical integration. The emission is always observed close to the electron's initial direction of propaga-

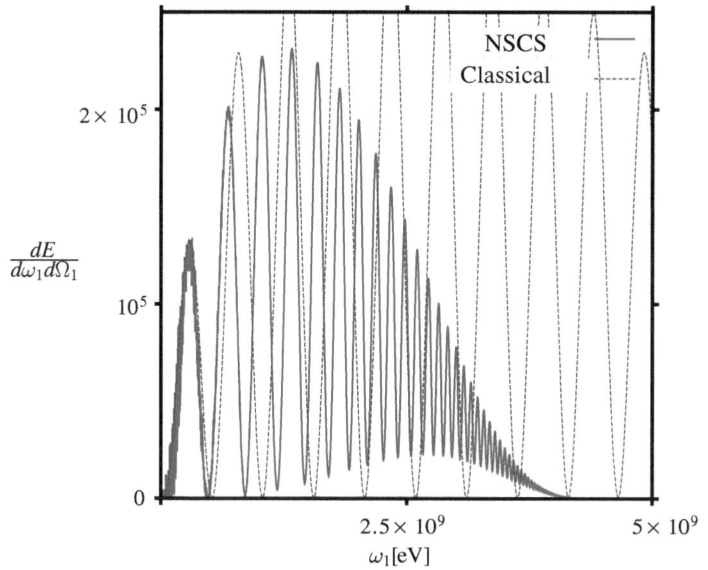

Fig. 3.3 Comparison of energy emission spectra, obtained via numerical integration of Eq. (3.21), to the classical result for the interaction of an electron with initial energy $\varepsilon_i = 5\,\text{GeV}$ colliding head on with a laser pulse described by Eq. (2.6) with $n_C = 2$ and intensity of $I_L = 2 \times 10^{20}\,{}^{\text{W}}/_{\text{cm}^2}$ ($\xi \approx 10$, $\chi \approx 0.6$). The spectra are given in units of [sr^{-1}]

tion $\vartheta_1 = \pi - \vartheta_0/2$. The angle $\vartheta_0 = {}^{m\xi}/_{\varepsilon_i}$ is shown to be of particular significance in Sect. 3.3. Indeed we find a perfect match for varying laser intensities, as long as $\chi \ll 1$ is satisfied. That for $\chi \sim 1$ the classical and the quantum result differ strongly is demonstrated in Fig. 3.3. In addition to the results of Eqs. (2.35) and (3.21) in Fig. 3.2 we have also marked the positions of the harmonic frequencies, obtained from a monochromatic analysis according to Eq. (2.125). In Fig. 3.2a we find differing, but still somewhat related results for the positions of the first two resonance peaks in the NSCS spectrum. In Fig 3.2b, on the other hand, we find that even the 10th monochromatic harmonic frequency significantly underruns the first resonance frequency of the NSCS result. We thus conclude that the monochromatic results are not suitable predict the photon frequencies emitted by an electron scattered from a few-cycles laser pulse, not to mention the width of the spectral resonance peaks. In Sect. D.2, however, we demonstrate that the monochromatic result in fact is reproduced, when considering a strictly periodic shape function in the dynamic integrals of Eq. (3.5). This result is equally evident from Fig. 3.4. There we show differential emission probabilities for an electron with initial energy $\varepsilon_i = 10$ MeV scattered from a laser pulse with intensity of $I = 5 \times 10^{18}\,{}^{\text{W}}/_{\text{cm}^2}$ (corresponding to an intensity parameter of $\xi \approx 1.5$) modeled by Eq. (2.8) for various numbers of switch-on cycles n_{switch}, contained in the leading and trailing edge of the pulse, and flat-top cycles n_{flat}. This shape function tends to a pure sine wave for $n_{\text{flat}} \to \infty$. We compute the spectra for an observation angle $\vartheta_1 = \pi$ and consider different num-

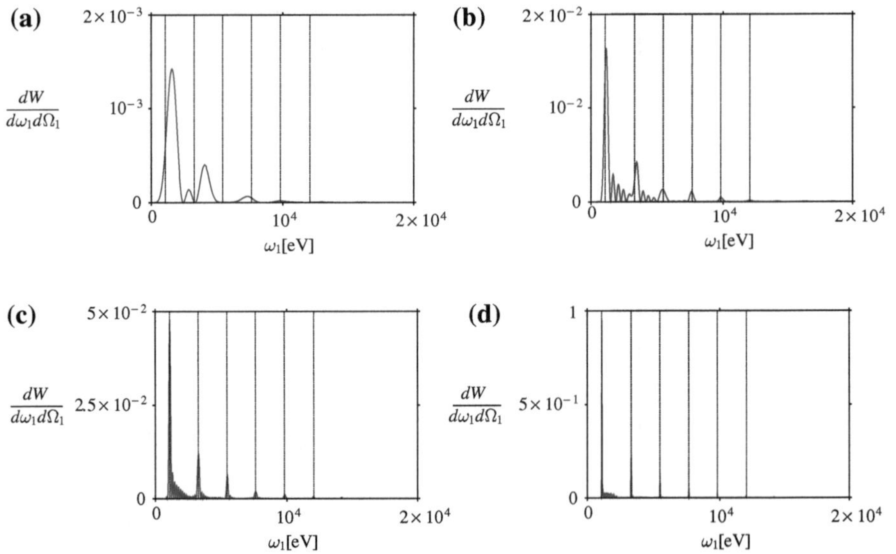

Fig. 3.4 NSCS emission probabilities for the shape function Eq. (2.8) with different numbers of contained cycles. The *dashed vertical lines* lie at the odd harmonics, obtained from analysis of a monochromatic laser pulse, according to Eq. (2.125). The probabilities are given in units of $[(\text{eV sr})^{-1}]$. **a** $n_{\text{flat}} = 0, n_{\text{switch}} = 1$. **b** $n_{\text{flat}} = 2, n_{\text{switch}} = 1$. **c** $n_{\text{flat}} = 6, n_{\text{switch}} = 2$. **d** $n_{\text{flat}} = 30, n_{\text{switch}} = 10$

bers of cycles contained in the pulses. We can clearly track the tendency towards the monochromatic emission frequencies, indicated by the dashed vertical lines, for $n_{\text{flat}} \to \infty$, alongside the increasing suppression of the side peaks lying in the vicinity of the main peaks.

The perturbative limit, recovered in Sect. D.1, is not backed up by numerical simulations of Eq. (3.21). We rather show how its reduces to the well studied limit of single photon Compton scattering [11–13] when one additionally assumes the laser field to be monochromatic. Analyzing the perturbative limit without this additional assumption, we find the emission probability to be given by a Fourier integral over the Compton scattering probabilities from the respective modes of the scattering laser pulse. This intuitive result has not been reported so far, to best of the author's knowledge.

As an important asymptotic limit, we next wish to explicitly demonstrate a highly useful stationary phase approximation to the dynamic integrals in the ultra-relativistic regime $\xi, \gamma \gg 1$. Given this realm of applicability this scheme is complementary to the previously mentioned low energy approximations.

3.3 Stationary Phase Approximation

Due to the general development of laser systems towards ever higher focused energy densities, the limiting case of highest interest for this thesis is the case of very large laser intensities $\xi \gg 1$. Thus we will work out the approximation technique for this case to a considerable degree. It will show, that the analytic results of this section intuitively result in possible applications. The following discussion is going to be technically rather involved. The main results of the stationary phase approximation, however, are summarized in a text passage at the end of this section and the reader is referred to this recapitulation for a concise overview.

In Chap. 1 we saw that a relativistic electron, except for the emission of radiation, can be described by means of a classical trajectory. We will thus lead the discussion with some intuitive classical considerations, deduced from the discussion in Sect. 2.1. There it was shown that an electron scattered from a laser pulse fulfilling $\xi \gg 1$ can be regarded as relativistic throughout the whole interaction, even if it was initially at rest. Since a relativistic electron radiates mostly along its instantaneous velocity (see Sect. 2.1), we expect the radiation to be emitted into those angular regions where the electron propagates into. This region is determined by the ratio $\varepsilon_i/m\xi$ (see Eq. (2.31b)), whence we expect radiation mainly into regions close to polar angles $\pi - \vartheta_1 \sim m\xi/\varepsilon_i$ for $\varepsilon_i \gg m\xi$, $\vartheta_1 \sim \varepsilon_i/m\xi$ for $m\xi \gg \varepsilon_i$ or essentially within the whole interval $\vartheta_1 \in [0, \pi]$ for $\varepsilon_i \sim m\xi$. Furthermore since the electron will only propagate in the laser's plane of polarization the emission outside of this plane is confined to azimuthal angles inside the instantaneous emission cone. The opening angle of this cone is proportional to the electron's instantaneous energy which, according to Eq. (2.31b), is given by $\varepsilon(\eta) \sim \varepsilon_i$ for $\varepsilon_i \gg m\xi$ and $\varepsilon(\eta) \sim (m\xi)^2/\varepsilon_i$ for $m\xi \gg \varepsilon_i$. The two scaling laws coincide for $m\xi \sim \varepsilon_i$ as they have to. Hence we conclude $|\varphi_1| \sim \min\{\varepsilon/(m\xi^2), m/\varepsilon\} \ll 1$. The same scaling holds equally for $|\pi - \varphi_1|$, which simply states that in the relativistic regime the electron radiates only in the plane of polarization. Inserting these scaling laws into Eqs. (3.8a)–(3.8c) for the Volkov parameters entering the exponential phase $S^{k_1}_{p_i,p_f}(x^-)$, we can determine the scaling properties of these parameters. We distinguish several scenarios where we first assume quantum recoil to be negligible $k_L k_1 \ll p_i k_L$ (see Eq. (1.7)). In this case we find

$$\alpha_V, \beta_V, \gamma_V \sim \xi^3. \tag{3.22}$$

If recoil has to be taken into account ($k_L k_1 \sim p_i k_L$), the only difference to the recoilless case is that it no longer holds $p_f^- \approx p_i^- \sim m\gamma$, but p_f^- is considerably reduced by k_1^-. Thus an additional factor $\varepsilon_i/p_f^- \gtrsim 1$ needs to be added for each factor. Thus for $\xi \gg 1$ the parameters α_V, β_V, γ_V scale equally in ξ and are all very large. Consequently the exponential phase in the dynamic integrals is very large and the integrals themselves are highly oscillating. This allows for an evaluation of the integrals by means of the stationary phase approximation. This method relies on the fact that integrals over rapidly oscillating functions of the form $\exp[if(x)]$, $|f(x)| \gg 1$, cancel themselves almost everywhere, except in the vicinity of points where $\partial_x f(x) = 0$. Such so-called points of stationary phase Ξ_0^-, are distinguished for the dynamic integrals Eq. (3.5) by the condition

$$\frac{\partial}{\partial x^-} S^{k_1}_{p_i,p_f}(x^-) = k_L^+ \left(\alpha_V \psi_{\mathcal{A}}(\Xi_0^-) + \beta_V \psi^2_{\mathcal{A}}(\Xi_0^-) + \gamma_V \right) = 0. \tag{3.23}$$

For $k_L^+ \neq 0$ this quadratic equation is solved by

$$\psi_{\mathcal{A}}(\Xi_0^-) = -\frac{\alpha_V}{2\beta_V} \pm \sqrt{\left(\frac{\alpha_V}{2\,\beta_V}\right)^2 - \frac{\gamma_V}{\beta_V}}. \tag{3.24}$$

For the important ratios of the exponential parameters from Eq. (3.8), contained in this equation, we find

$$\frac{\alpha_V}{2\,\beta_V} = \frac{e}{m^2\xi^2} \frac{(p_f A_L)(p_i k_L) - (p_i A_L)(k_L p_f)}{k_L k_1} \tag{3.25a}$$

$$\frac{\gamma_V}{\beta_V} = -\frac{2(k_1 p_i)(p_i k_L)}{m^2\xi^2(k_L k_1)} \tag{3.25b}$$

It was previously shown that the determinant of the square root in Eq. (3.24) is always imaginary and suppressed by ξ^{-1} [10]. We wish to briefly sketch the central points of the calculation. By virtue of Eqs. (3.25) and the energy momentum conservation contained in Eq. (3.3) the determinant can be written in the form

$$\begin{aligned}\left(\frac{\alpha_V}{2\,\beta_V}\right)^2 - \frac{\gamma_V}{\beta_V} &= (m\xi)^{-2} \left[\left(\epsilon_L \boldsymbol{p}_i - \epsilon_L \boldsymbol{k}_1 \frac{p_i^-}{k_1^-} \right)^2 - 2(k_1 p_i) \frac{p_i^-}{k_1^-} \right] \\ &= -(m\xi)^{-2} \left[\left(k_1^\perp \left(\frac{p_i^-}{k_1^-} \right) - p_i^\perp \right)^2 + m^2 \right].\end{aligned} \tag{3.26}$$

We consequently find the previous claim on the properties of Eq. (3.24) confirmed. Translating Eq. (3.24) to a solution for the stationary point Ξ_0^- itself, we find an imaginary contribution also suppressed as ξ^{-1}. Due to its suppression, we can treat the imaginary part of Eq. (3.24) as a small perturbation to the full stationary point in the sense

$$\Xi_0^- = \psi_{\mathcal{A}}^{-1} \left(\frac{\alpha_V}{2\beta_V} + \mathrm{i}\kappa \right) \tag{3.27}$$

$$\kappa = \pm \sqrt{\frac{\gamma_V}{\beta_V} - \left(\frac{\alpha_V}{2\,\beta_V}\right)^2}. \tag{3.28}$$

The expression for Ξ_0^- can then be expanded around $\kappa = 0$. We comment on the correct choice of the sign of κ further down. The sketched procedure results in

$$\Xi_0^- = \mathring{x}^- + \frac{\partial}{\partial\kappa}\Xi_0^-\bigg|_{\kappa=0}\kappa + \mathcal{O}(\kappa^2) \tag{3.29}$$

where we defined the real part of the stationary point $\mathring{x}^- = \psi_{\mathcal{A}}^{-1}\left(-\frac{\alpha_V}{2\beta_V}\right)$. The symbol is chosen to make a clear distinction to the carrier-envelope phase $x_0^- = \eta_0/k_L^+$. According to the inverse function theorem the differential with respect to κ is evaluated as

$$\frac{\partial}{\partial\kappa}\Xi_0^-\bigg|_{\kappa=0} = \frac{\partial\psi_{\mathcal{A}}(\Xi_0^-)}{\partial\kappa}\frac{\partial}{\partial\psi_{\mathcal{A}}(\Xi_0^-)}\psi_{\mathcal{A}}^{-1}(\psi_{\mathcal{A}}(\Xi_0^-))\bigg|_{\kappa=0} = \frac{\mathrm{i}}{\psi'_{\mathcal{A}}(\mathring{x})}. \tag{3.30}$$

In the latter two expressions $\psi_{\mathcal{A}}(\mathring{x}^-)$ is purely real and the derivative $\psi'_{\mathcal{A}}(\mathring{x}^-)$ is of order unity for an oscillating function. The stationary point Ξ_0^- consequently is expressible as

$$\Xi_0^- \approx \mathring{x}^- + \frac{\mathrm{i}}{\psi'_{\mathcal{A}}(\mathring{x}^-)}\kappa \approx \mathring{x}^- = \psi_{\mathcal{A}}^{-1}\left(-\frac{\alpha_V}{2\beta_V}\right). \tag{3.31}$$

From now on for the remainder of this chapter, in all expressions depending on the stationary point we will denote this dependency by either $\ldots(\Xi_0^-)$ or $\ldots(\mathring{x}^-)$, depending on whether the imaginary part in Eqs. (3.24) and (3.31) has to be taken into account or not, respectively. The real part $\mathring{x}^-$ itself can now be real only if the following condition is met

$$\psi_{\mathcal{A}}^{\min} \leq -\frac{\alpha}{2\beta} \leq \psi_{\mathcal{A}}^{\max}. \tag{3.32}$$

Here $\psi_{\mathcal{A}}^{\mathrm{min/max}}$ denote the minimal and maximal value the shape function can take for real x^-, respectively. Condition (3.32) is fundamentally different from Eq. (3.31), where we saw that $\partial_{x^-}S_{p_i,p_f}^{k_1}(x^-)$ never vanishes on the real axis, but the imaginary contribution to a stationary point Ξ_0^- is small in the regime $\xi \gg 1$. On the other hand, if condition (3.32) is not fulfilled then $\mathring{x}^-$ contains an imaginary part of order unity even if κ is negligible. Equation (3.32) can then be employed to deduce a relation between the spatial regions where an electron is expected to emit radiation and the specific form of the shape function (see Sect. 3.4 and [10, 14]). If we consider only such situations in which the condition (3.32) is fulfilled, the functions f_i can be evaluated by applying a stationary phase approximation to the integrals in x^-. Following this method, the exponential as well as the preexponentials in the dynamic integrals (3.5) may be expanded in a perturbation series in $\left(x^- - \mathring{x}^-\right)$ since for values of x^- far away from the stationary point, the rapid oscillations of the integrand will cancel themselves. This expansion is of the form

$$f_i \doteq \int_{-\infty}^{\infty} dx^- \sum_{n=0}^{N} \frac{(x^- - \mathring{x}^-)^n}{n!} \left(\frac{\partial}{\partial x^-}\right)^n \left[\psi^i_{\mathcal{A}}(x^-)\right]_{x^-=\mathring{x}^-}$$
$$\times \exp\left[-\mathrm{i} \sum_{m=0}^{M} \frac{(x^- - \mathring{x}^-)^m}{m!} \partial^m_{x^-} S^{k_1}_{p_i,p_f}(x^-)\Big|_{x^-=\Xi_0^-} \right]. \tag{3.33}$$

Here $S^{k_1}_{p_i,p_f}(x^-)$ is the classical action from Eq. (3.12) and N and M are the orders up to which the preexponential and the exponential functions are expanded, respectively. In Eq. (3.33) we kept the exact stationary point Ξ_0^- in the exponential, since including only its real part $\mathring{x}^-$ leads to cancellations, as we will see. Inserting the lowest order expansion in the preexponential for $N = 0$ into the squared matrix element, we also find a cancellation for the preexponential

$$|S_{fi}|^2 \approx \left| -\left(\tilde{\Gamma}_1 - \frac{\alpha_V}{\gamma_V}\not{\epsilon}_1^*\right) \frac{\alpha_V}{2\beta_V}\mathcal{I}_0 + \left(\tilde{\Gamma}_2 - \frac{\beta_V}{\gamma_V}\not{\epsilon}_1^*\right)\left(\frac{\alpha_V}{2\beta_V}\right)^2 \mathcal{I}_0 \right|^2$$
$$= \left| \tilde{\Gamma}_2 \left(\frac{\alpha_V}{2\beta_V}\right)^2 - \tilde{\Gamma}_1 \frac{\alpha_V}{2\beta_V} \right|^2 |\mathcal{I}_0|^2 . \tag{3.34}$$

Here $\mathcal{I}_0$ is the first integral of the preexponential series in Eq. (3.33), explicitly given below. Due to this apparent cancellation, the preexponential perturbation series needs to be taken into account up to next-to-leading order at least. The described cancellations in the exponential are even stronger. Taking the derivative of the action $S(x^-)$ we find

$$\partial_{x^-} S^{k_1}_{p_i,p_f}(\Xi_0^-) = 0 \qquad \text{(by definition)} \tag{3.35a}$$
$$\partial^2_{x^-} S^{k_1}_{p_i,p_f}(\Xi_0^-) = k_L^+ \left(\alpha_V \psi'_{\mathcal{A}}(\Xi_0^-) + 2\beta_V\, \psi'_{\mathcal{A}}(\Xi_0^-)\psi_{\mathcal{A}}(\Xi_0^-)\right), \tag{3.35b}$$
$$\partial^3_{x^-} S^{k_1}_{p_i,p_f}(\Xi_0^-) = k_L^+ \left(\alpha_V \psi''_{\mathcal{A}}(\Xi_0^-) + 2\beta_V \left(\psi''_{\mathcal{A}}(\Xi_0^-)\psi_{\mathcal{A}}(\Xi_0^-) + \psi'^2_{\mathcal{A}}(\Xi_0^-)\right)\right), \tag{3.35c}$$

where $\psi'_{\mathcal{A}}(x^-)$ means the derivative with respect to x^-. For the sake of notational simplicity from now on we write $\partial^2_{x^-} S^{k_1}_{p_i,p_f}(\Xi_0^-) = \partial^2_{x^-} S^{k_1}_{p_i,p_f}(\Xi_0^-)$ and $\partial^3_{x^-} S^{k_1}_{p_i,p_f}(\mathring{x}^-) = \partial^3_{x^-} S^{k_1}_{p_i,p_f}(\Xi_0^-)$. If approximation Eq. (3.31) held exactly, $\partial^2_{x^-} S^{k_1}_{p_i,p_f}(\Xi_0^-) \equiv 0$ would follow. We thus conclude that the imaginary part of Ξ_0^- in this case is non-negligible. However, this leads to $\partial^2_{x^-} S^{k_1}_{p_i,p_f}(\Xi_0^-) = 2k_L^+ \beta_V \psi'_{\mathcal{A}}(\mathring{x}^-)\,\mathrm{i}\kappa \sim \xi^2$. Since in the expansion of Eq. (3.33) the exponential leads to oscillations only if it exceeds order unity, the integral over x^- will be non-negligible in a vicinity of $\mathring{x}^-$ of the size $(x^- - \mathring{x}^-) \sim \xi^{-1}$, since then $\partial^2_{x^-} S^{k_1}_{p_i,p_f}(\Xi_0^-)(x^- - \mathring{x}^-)^2 \sim 1$. The third order derivative, given in Eq. (3.35c), on the other hand, scales as $\partial^3_{x^-} S^{k_1}_{p_i,p_f}(\mathring{x}^-) = 2k_L^+ \beta_V \psi'^2_{\mathcal{A}}(\mathring{x}^-) \sim \xi^3$, where the imaginary part of Ξ_0^- is negligible again. It consequently holds $\partial^2_{x^-} S^{k_1}_{p_i,p_f}(\Xi_0^-) = \mathrm{i}\kappa/\psi'_{\mathcal{A}}(\mathring{x}^-)\partial^3_{x^-} S^{k_1}_{p_i,p_f}(\mathring{x}^-)$ and we conclude $\partial^3_{x^-} S^{k_1}_{p_i,p_f}(\mathring{x}^-)(x^- - \mathring{x}^-)^3 \sim \partial^2_{x^-} S^{k_1}_{p_i,p_f}(\Xi_0^l)(x^- - \mathring{x}^-)^2 \sim 1$. Thus both contributions need to be included in the exponential series in Eq. (3.33). We finally note, that, since Eq. (3.31) may have several solutions for an oscillating $\psi_{\mathcal{A}}(x^-)$, there may be more than one stationary point. In this case one has to sum over contributions from each stationary point $\Xi_0^{-,l}, \mathring{x}^{-,l}$ with the index l running over all solutions of Eq. (3.31). The expansion of the dynamic integrals is then of the general form

$$f_i \sim \sum_l \psi^i_{\mathcal{A}}(\mathring{x}^{-,l})\mathcal{I}^l_0 + \psi^{i\prime}_{\mathcal{A}}(\mathring{x}^{-,l})\mathcal{I}^l_1 + \frac{1}{2}\psi^{i\prime\prime}_{\mathcal{A}}(\mathring{x}^{-,l})\mathcal{I}^l_2. \tag{3.36}$$

The integrals in Eq. (3.36) are given by the definitions

$$\mathcal{I}^l_i = \int_{-\infty}^{\infty} dx^- \, (x^- - \mathring{x}^{-,l})^i \exp\left[-\mathrm{i}\Sigma^{k_1}_{p_i,p_f}(x^-)\right] \tag{3.37}$$

$$\Sigma^{k_1}_{p_i,p_f}(x^-) = \sum_{m=0}^{3} \frac{(x^- - \mathring{x}^{-,l})^m}{m!} \partial^m_{x^-} S^{k_1}_{p_i,p_f}(x^-)\bigg|_{x^- = \Xi_0^{-,l}}. \tag{3.38}$$

Because the suppression of the stationary points' imaginary part $\kappa \sim \xi^{-1}$ and the correction terms $(x^- - \mathring{x}^{-,l}) \sim \xi^{-1}$ we conclude that, approximating the dynamic integrals by Eqs. (3.36) and (3.37), will result in better approximations to the f_i, the larger the parameter ξ is. From Eq. (3.37) we can furthermore deduce the proper choice of the sign of the imaginary contribution κ^l, defined in Eq. (3.28). The exponential in Eq. (3.37) will contain a factor $-\mathrm{i}\partial^2_{x^-} S^{k_1}_{p_i,p_f}(\Xi_0^{-,l}) = 2k_L^+ \beta_V \psi'_{\mathcal{A}}(\mathring{x}^{-,l})\kappa^l \sim \xi^2$. This expression must be smaller than 0, for otherwise the integrals $\mathcal{I}^l_0$ are exponentially diverging. Recalling $\beta_V < 0$ [see Eq. (3.8b)], we infer as the proper choice of the sign of κ^l for any stationary point $\mathring{x}^{-,l}$

$$\operatorname{sgn}\left(\kappa^l\right) = \operatorname{sgn}\left(\psi'_{\mathcal{A}}(\mathring{x}^{-,l})\right). \tag{3.39}$$

Equation (3.37) needs to be evaluated explicitly only for the case $i = 0$. The cases $i = 1, 2$ can be derived from this result. The expression for $\mathcal{I}^l_0$ can be simplified using the changed integration variable $y_l = x^- - \mathring{x}^{-,l} + c_l$. Defining $c_l = -\partial^2_{x^-} S^{k_1}_{p_i,p_f}(\Xi_0^{-,l})/\partial^3_{x^-} S^{k_1}_{p_i,p_f}(\mathring{x}^{-,l})$, all terms quadratic in y_l drop out of Eq. (3.37) and the equation simplifies to

$$
\mathcal{I}_0^l = \exp\left[-\mathrm{i}\left(S(\Xi_0^{-,l}) + \frac{\left(\partial_{x^-}^2 S_{p_i,p_f}^{k_1}(\Xi_0^{-,l})\right)^3}{3\left(\partial_{x^-}^3 S_{p_i,p_f}^{k_1}(\mathring{x}^{-,l})\right)^2}\right)\right]
\times \int_{-\infty}^{\infty} dy_l \exp\left[-\mathrm{i}\left(y_l^3 \frac{\partial_{x^-}^3 S_{p_i,p_f}^{k_1}(\mathring{x}^{-,l})}{6} - y_l \frac{\left(\partial_{x^-}^2 S_{p_i,p_f}^{k_1}(\Xi_0^{-,l})\right)^2}{2\partial_{x^-}^3 S_{p_i,p_f}^{k_1}(\mathring{x}^{-,l})}\right)\right]. \quad (3.40)
$$

Along the lines of the argumentation leading to Eq. (3.29), we expand the stationary phase $S(\Xi_0^{-,l})$ around its real value $S(\mathring{x}^{-,l})$ and treat the imaginary contribution $\mathrm{i}\kappa$ as a small perturbation. From this procedure we read off from Eq. (3.31) $\partial^n/\partial\kappa^n S(\Xi_0^{-,l}) = \mathrm{i}^n/\psi_{\mathcal{A}}'^{n}(\mathring{x}^{-,l}) \left(\partial/\partial\Xi_0^{-,l}\right)^n S(\Xi_0^{-,l})$, whence we deduce

$$
S_{p_i,p_f}^{k_1}(\Xi_0^{-,l}) \approx S_{p_i,p_f}^{k_1}\left(\mathring{x}^{-,l}\right) - \frac{\partial_{x^-}^2 S_{p_i,p_f}^{k_1}(\Xi_0^{-,l})}{\psi_{\mathcal{A}}'^{2}(\mathring{x}^{-,l})}\frac{\kappa^{l2}}{2} - \frac{\mathrm{i}\partial_{x^-}^3 S_{p_i,p_f}^{k_1}(\mathring{x}^{-,l})}{\psi_{\mathcal{A}}'^{3}(\mathring{x}^{-,l})}\frac{\kappa^{l3}}{6}. \quad (3.41)
$$

Recall Eqs. (3.35) we conclude that the imaginary part of $S(\Xi_0^{-,l})$ thus is given by the sum of the quadratic and the cubic term in Eq. (3.41) which are both of order $\beta_V\kappa^3 \sim 1$. From Eqs. (3.35b) and (3.35c) we read off $\mathrm{Im}(S(\Xi_0^{-,l})) = \mathrm{i}\beta_V k_L^+\kappa^{l3}/\psi'_{\mathcal{A}}(\mathring{x}^{-,l}) - \mathrm{i}\beta_V k_L^+\kappa^{l3}/3\psi'_{\mathcal{A}}(\mathring{x}^{-,l}) = 2\mathrm{i}\beta_V k_L^+\kappa^{l3}/3\psi'_{\mathcal{A}}(\mathring{x}^{-,l})$. Additionally $(\partial_{x^-}^2 S_{p_i,p_f}^{k_1}(\Xi_0^{-,l}))^3/3\left(\partial_{x^-}^3 S_{p_i,p_f}^{k_1}(\mathring{x}^{-,l})\right)^2 = -2\mathrm{i}\beta_V k_L^+\kappa^{l3}/3\psi'_{\mathcal{A}}(\mathring{x}^{-,l})$ is found from Eq. (3.35) and the subsequent discussion. For the constant exponential factor in front of the integral in Eq. (3.40) we thus find

$$
S_{p_i,p_f}^{k_1}(\Xi_0^{-,l}) + \frac{\left(\partial_{x^-}^2 S_{p_i,p_f}^{k_1}(\Xi_0^{-,l})\right)^3}{3\left(\partial_{x^-}^3 S_{p_i,p_f}^{k_1}(\mathring{x}^{-,l})\right)^2} = S_{p_i,p_f}^{k_1}\left(\mathring{x}^{-,l}\right). \quad (3.42)
$$

The constant phase in front of the integral in Eq. (3.40) thus is purely imaginary and leads to a rapid oscillation of the overall integral $\mathcal{I}_0^l$. The remaining integration over y_l in that equation is elementary solvable in terms of the Airy function $\mathrm{Ai}(x)$ [15] and the result is

$$
\mathcal{I}_0^l \sim 2\,\mathrm{e}^{\mathrm{i}S_{p_i,p_f}^{k_1}(\mathring{x}^{-,l})}\left(\frac{2\pi^3}{\partial_{x^-}^3 S_{p_i,p_f}^{k_1}(\mathring{x}^{-,l})}\right)^{\frac{1}{3}} \mathrm{Ai}(\lambda_l)
= 2\,\mathrm{e}^{\mathrm{i}S_{p_i,p_f}^{k_1}(\mathring{x}^{-,l})}\left(\frac{\pi^3}{|\beta_V|k_L^+\psi_{\mathcal{A}}'^{2}(\mathring{x}^{-,l})}\right)^{\frac{1}{3}} \mathrm{Ai}(\lambda_l). \quad (3.43)
$$

By employing the relation between $\partial^2_{x^-} S^{k_1}_{p_i,p_f}(\Xi_0^{-,l})$ and $\partial^3_{x^-} S^{k_1}_{p_i,p_f}(\mathring{x}^{-,l})$ stated above, the argument λ_l can be written as

$$\lambda_l = \frac{S^{k_1\,\prime\prime\,2}_{p_i,p_f}(\Xi_0^{-,l})}{\left(4\left(\partial^3_{x^-} S^{k_1}_{p_i,p_f}(\mathring{x}^{-,l})\right)^4\right)^{1/3}} = \left(\frac{\kappa^{l\,3} k_L^+ |\beta_V|}{\psi'_{\mathcal{A}}(\mathring{x}^{-,l})}\right)^{\frac{2}{3}}. \tag{3.44}$$

With an explicit expression for the integral $\mathcal{I}_0^l$ it is possible to derive an expression for the integral $\mathcal{I}_2^l$. To this end we note that from Eq. (3.37) it follows

$$\mathcal{I}_2^l = 2\mathrm{i}\,\frac{\partial}{\partial(\partial^2_{x^-} S^{k_1}_{p_i,p_f}(\Xi_0^{-,l}))}\mathcal{I}_0^l. \tag{3.45}$$

The power law prefactor in front of the Airy function in Eq. (3.43) is independent of, whereas the exponential phase factor depends on $\partial^2_{x^-} S^{k_1}_{p_i,p_f}(\Xi_0^{-,l})$ as can be seen from Eq. (3.40). The explicit derivatives of these two terms are given by

$$\frac{\partial}{\partial\partial^2_{x^-} S^{k_1}_{p_i,p_f}(\Xi_0^{-,l})}\,\mathrm{e}^{-\mathrm{i}S^{k_1}_{p_i,p_f}(\mathring{x}^{-,l})} = \frac{2\lambda_l^{\frac{3}{2}}}{\partial^2_{x^-} S^{k_1}_{p_i,p_f}(\Xi_0^{-,l})}\,\mathrm{e}^{-\mathrm{i}S^{k_1}_{p_i,p_f}(\mathring{x}^{-,l})} \tag{3.46a}$$

$$\frac{\partial}{\partial\partial^2_{x^-} S^{k_1}_{p_i,p_f}(\Xi_0^{-,l})}\lambda_l = \frac{2\lambda_l}{\partial^2_{x^-} S^{k_1}_{p_i,p_f}(\Xi_0^{-,l})}, \tag{3.46b}$$

where in Eq. (3.46a) we had to reinsert Eq. (3.42) for $S^{k_1}_{p_i,p_f}\left(\mathring{x}^{-,l}\right)$. It is then straightforward to obtain an explicit representation of the integral function $\mathcal{I}_2^l$ in the following form

$$\mathcal{I}_2^l \approx \frac{8\mathrm{e}^{\mathrm{i}\partial^2_{x^-} S^{k_1}_{p_i,p_f}(\mathring{x}^{-,l})}}{(-\mathrm{i}\partial^2_{x^-} S^{k_1}_{p_i,p_f}(\Xi_0^{-,l}))}\left(\frac{\pi^3}{|\beta_V| k_L^+ \psi'^2_{\mathcal{A}}(\mathring{x}^{-,l})}\right)^{\frac{1}{3}}\left(\lambda_l^{\frac{3}{2}}\,\mathrm{Ai}\left(\lambda_l\right) + \lambda_l\,\mathrm{Ai}'\left(\lambda_l\right)\right). \tag{3.47}$$

For the computation of the last integral $\mathcal{I}_1^l$, we employ an integration by parts technique in analogy to Eq. (3.13) and obtain

$$\mathcal{I}_1^l \approx -\frac{\partial^3_{x^-} S^{k_1}_{p_i,p_f}(\mathring{x}^{-,l})}{2\partial^2_{x^-} S^{k_1}_{p_i,p_f}(\Xi_0^{-,l})}\mathcal{I}_2^l. \tag{3.48}$$

Here in the transition from the first to the second line we already dropped the contributions at $\pm\infty$. The Eqs. (3.43), (3.47) and (3.48) inserted into Eq. (3.36) provide an asymptotic expansion of the dynamic integrals f_i in the ultra relativistic regime. We note that the dynamic integrals can accordingly be expressed as

$$f_i = \sum_{l=1}^{N_{\text{s.p.}}} f_i^{(l)}, \tag{3.49}$$

where $f_i^{(l)}$ denotes the contribution from the lth stationary point $\mathring{x}^{-,l}$ and $N_{\text{s.p.}}$ gives the total number of stationary points. This sum expansion translates to the overall expression of the matrix element Eq. (3.2), which then reads

$$S_{fi} = \sum_{l=1}^{N_{\text{s.p.}}} S_{fi}^{(l)}. \tag{3.50}$$

The just presented stationary phase approximation has a beautiful physical visualization, namely it allows for a quasi-classical interpretation of the electron emission. To demonstrate this we will show that the found stationary points correspond to the points in phase $\eta^l = k_L^+ \mathring{x}^{-,l}$ where the velocity vector of a classical electron points into the chosen direction of observation. Since the electron is assumed to be relativistic, these are also the points of the trajectory, where it emits into the observation direction. We thus have to compute the angles (ϑ_1, φ_1), into which a classical electron's velocity points at the segments of its trajectory, distinguished by the condition $\psi_{\mathcal{A}}(x^-) = \psi_{\mathcal{A}}(\mathring{x}^-)$. We note that this condition is the same for all stationary points, whence the superscript l can be omitted. Since the motion of a classical electron in a linearly polarized plane wave is confined to the wave's plane of polarization (recall Fig. 2.2a) we can restrict ourselves to $\varphi_1 = 0, \pi$ and compute the corresponding value of ϑ_1 only. In fact, in the following we will only sketch the computation for $\varphi_1 = 0$, corresponding to $\beta^x(\mathring{x}^-) > 0$, since the opposite case $\varphi_1 = \pi$ is analogous. From Eq. (2.31b), evaluated in the coordinate frame specified in Fig. 2.1, we obtain in the relativistic regime $m\xi, \varepsilon_i \gg m$

$$\begin{aligned} \tan(\vartheta_1^{\text{class}}) &= \frac{\beta^x(\mathring{x}^-)}{\beta^z(\mathring{x}^-)} = \frac{m\xi\psi_{\mathcal{A}}(\mathring{x}^-)}{p_i^z - \omega_L \frac{m^2\xi^2}{2(p_i k_L)}\psi_{\mathcal{A}}^2(\mathring{x}^-)} \\ &\approx \frac{k_1^x}{\frac{k_1^-}{\sqrt{2}} + \frac{k_1^+}{\sqrt{2}}} = \frac{k_1^x}{k_1^z} \\ &= \tan(\vartheta_1) . \end{aligned} \tag{3.51}$$

Since for $\vartheta_1^{\text{class}}, \vartheta_1 \in [0, \pi]$ Eq. (3.51) is solved only for $\vartheta_1^{\text{class}} = \vartheta_1$ this proves the conjecture. The explanation for this quasi-classical behaviour was given in [2] in terms of the so-called *coherence interval*. This notion refers to the length of a trajectory segment, over which the radiation from an electron is coherent, and which thus has to be taken into account in analyzing a quantum process depicted in Fig. 3.1.

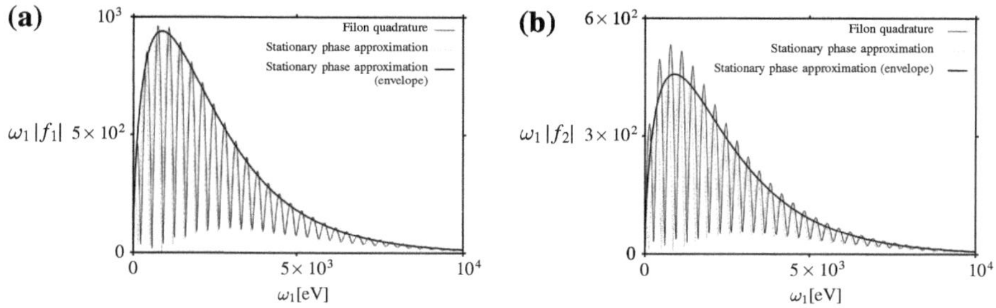

Fig. 3.5 Dynamic integrals obtained from the stationary phase approximation for an electron of initial energy $\varepsilon_i = 10\,\text{MeV}$ scattered from a laser pulse $I_L = 10^{20}\,\text{W/cm}^2$ observed at $(\vartheta_1, \varphi_1) = (170°, 0°)$. **a** First dynamic integral. **b** Second dynamic integral

It was shown that in the relativistic regime the coherence interval stretches over a phase interval interval of the length $\Delta\eta \sim \xi^{-1} \ll 1$. One can thus consider each emission process to be formed at a single point of the trajectory, whence detecting single photons will probe a quasi-classical trajectory of the electron.

To finally demonstrate the numerical exactness and thus predictive power of the demonstrated stationary phase approximation, we will consider two numerical examples. As a first example we show in Fig. 3.5 the stationary phase approximations to $\omega_1 |f_{1,2}|$, obtained from Eqs. (3.36), (3.43), (3.47) and (3.48), in comparison to an exact numerical quadrature of the dynamic integrals (3.5), presented in Chap. 5. Recall that the function f_0 is found accordingly by employing Eq. (3.13). The dynamic integrals are computed for the scattering of an electron with initial energy of $\varepsilon_i = 10\,\text{MeV}$ from a laser pulse described by Eq. (2.6) with $n_C = 2$ and a moderate intensity of $I = 10^{20}\,\text{W/cm}^2$ ($\xi \approx 7$) observed at the direction $\vartheta_1 = \pi - \vartheta_0/2 \approx 175°$ and $\varphi_1 = 0$ in a reference frame according to Fig. 2.1. We show the complete approximation, as well as the envelopes of these functions, which are obtained by omitting the oscillating phase term $\exp\left[\mathrm{i}S^{k_1}_{p_i,p_f}\left(\mathring{x}^{-,l}\right)\right]$ from the definitions of the $\mathcal{I}^l_i$. Comparing the stationary phase approximation with the exact result, we observe a fine agreement already for moderate values of $\xi \sim 1-10$. Since for higher values of ξ the oscillations in the dynamic integrals are going to increase strongly (recall that the oscillations reflect the absorption of higher numbers of photons and at $\xi \gg 1$ the electron interacts with $\sim \xi^3$ photons from the laser field), in the following examples we are going to show merely the envelope of the stationary phase approximations to the dynamic integrals. Increasing the considered values of the classical intensity parameter, this agreement is expected to further improve. And in fact, comparing the exact quadrature with the results of the stationary phase approximation for an electron with initial energy of $\varepsilon_i = 100\,\text{MeV}$ scattered from a laser pulse with intensity of $I = 10^{22}\,\text{W/cm}^2$ ($\xi \approx 70$) (again observed at the direction ($\vartheta_1 = \pi - \vartheta_0/2 \approx 170°, \varphi_1 = 0$)), shown in Fig. 3.6, we find an agreement which cannot be told apart by the eye. As announced before in these spectra we only show the envelope of the dynamic integrals obtained via the stationary phase

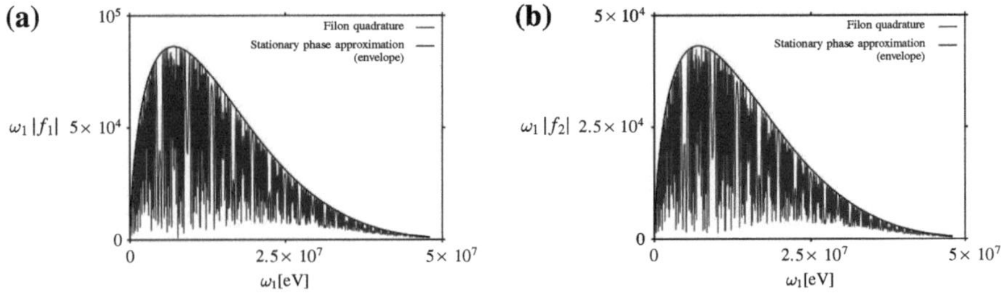

Fig. 3.6 Dynamic integrals obtained from the stationary phase approximation for an electron of initial energy $\varepsilon_i = 100\,\mathrm{MeV}$ scattered from a laser pulse $I_L = 10^{22}\,\mathrm{W/cm^2}$ observed at $(\vartheta_1, \varphi_1) = (170°, 0°)$. **a** First dynamic integral. **b** Second dynamic integral

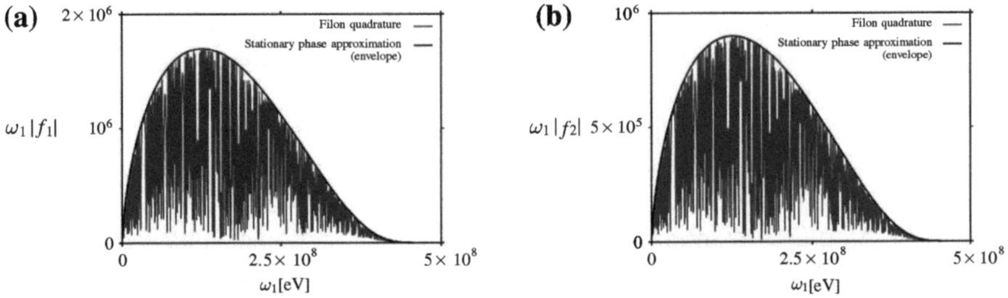

Fig. 3.7 Dynamic integrals obtained from the stationary phase approximation for an electron of initial energy $\varepsilon_i = 500\,\mathrm{MeV}$ scattered from a laser pulse $I_L = 2\times 10^{22}\,\mathrm{W/cm^2}$ observed at $(\vartheta_1, \varphi_1) = (177°, 0°)$. **a** First dynamic integral. **b** Second dynamic integral

approximation. Furthermore from the parameters, these figures are computed for, we derive a quantum nonlinearity parameter of $\chi \approx 0.1$. To investigate even further pronounced quantum effects, we finally study a set of experimental parameters aiming at the frontier of the nowadays available technical potential. We consider an electron with initial energy of $\varepsilon_i = 500\,\mathrm{MeV}$ scattered from a laser pulse with intensity of $I = 2\times 10^{22}\,\mathrm{W/cm^2}$ ($\xi \approx 100$) ($\vartheta_1 = \pi - \vartheta_0/2 \approx 177°, \varphi_1 = 0$) and show the resulting dynamic integrals in Fig. 3.7. In these figures we find a significant distortion of the spectra's shapes, which is due to the suppression of high photon frequencies close to the quantum mechanical cutoff energy, distinguished by Eq. (3.16). So we conclude that, as it has to be, also for quantum effects becoming increasingly important, the stationary phase method provides us with a reliable approximation for the dynamic integrals. Of course, reducing the intensity parameter further, or even below $\xi \sim 1$, will decrease and eventually completely abolish the predictive power of the presented method.

3.3.1 Summarizing the Above Discussion

In the regime $m\xi$, $\varepsilon_i \gg m$ the dynamic integrals f_i from Eq. (3.5) can be approximated by

$$f_i \approx \sum_l \psi^i_{\mathcal{A}}(\mathring{x}^{-,l})\mathcal{I}^l_0 + \psi^{i\,\prime}_{\mathcal{A}}(\mathring{x}^{-,l})\mathcal{I}^l_1 + \frac{1}{2}\psi^{i\,\prime\prime}_{\mathcal{A}}(\mathring{x}^{-,l})\mathcal{I}^l_2$$

$$\mathcal{I}^l_0 = 2\,\mathrm{e}^{\mathrm{i}S^{k_1}_{p_i,p_f}(\mathring{x}^{-,l})}\left(\frac{\pi^3}{|\beta_V|k_L^+\psi'^2_{\mathcal{A}}(\mathring{x}^{-,l})}\right)^{\frac{1}{3}}\mathrm{Ai}\,(\lambda_l)$$

$$\mathcal{I}^l_1 = -\frac{\partial^3_{x^-}S^{k_1}_{p_i,p_f}(\mathring{x}^{-,l})}{2\partial^2_{x^-}S^{k_1}_{p_i,p_f}(\Xi^{-,l}_0)}\mathcal{I}^l_2$$

$$\mathcal{I}^l_2 = \frac{8\mathrm{e}^{\mathrm{i}\partial^2_{x^-}S^{k_1}_{p_i,p_f}(\mathring{x}^{-,l})}}{(-\mathrm{i}\partial^2_{x^-}S^{k_1}_{p_i,p_f}(\Xi^{-,l}_0))}\left(\frac{\pi^3}{|\beta_V|k_L^+\psi'^2_{\mathcal{A}}(\mathring{x}^{-,l})}\right)^{\frac{1}{3}}\left(\lambda_l^{\frac{3}{2}}\,\mathrm{Ai}\,(\lambda_l) + \lambda_l\,\mathrm{Ai}'\,(\lambda_l)\right)$$

$$\Xi^{-,l}_0 = \mathring{x}^{-,l} + \frac{\mathrm{i}}{\psi'_{\mathcal{A}}(\mathring{x}^{-,l})}\kappa$$

$$\mathring{x}^{-,l} = \psi^{-1}_{\mathcal{A}}\left(-\frac{\alpha_V}{2\beta_V}\right), \qquad \kappa^l = \mathrm{sgn}\left(\psi'_{\mathcal{A}}(\mathring{x}^{-,l})\right)\sqrt{\frac{\gamma_V}{\beta_V} - \left(\frac{\alpha_V}{2\,\beta_V}\right)^2}.$$

The validity of these expressions is shown by comparing them to exact numerical quadratures of the f_i, to be found in Figs. 3.5, 3.6, 3.7. The exact quadrature scheme is outlined in Chap. 5. We furthermore found that, for any stationary point to exist at all, we must choose an observation direction $\boldsymbol{n}_1$, satisfying

$$\frac{(m\xi)}{\varepsilon_i}\psi^{\min}_{\mathcal{A}} \leq (\beta_i\epsilon_L) - (n_L\beta_i)\frac{\boldsymbol{k}_1\epsilon_L}{1-\boldsymbol{n}_L\boldsymbol{n}_1} \leq \frac{(m\xi)}{\varepsilon_i}\psi^{\max}_{\mathcal{A}}, \tag{3.52}$$

which is a recapture of Eq. (3.32) by means of Eq. (3.25a) and where $\beta^\mu_i = p^\mu_i/\varepsilon_i$ is the electron's initial four velocity (see Sect. 2.1). We finally found that the stationary points $\mathring{x}^{-,l}$ correspond to those points where a classical electron emits into the direction $\boldsymbol{n}_1$. By virtue of this correspondence we can establish a quasi-classical picture of the electron's dynamic inside the plane laser field. The angular distribution of the emission from the electron can be understood as being emitted by an electron following a classical trajectory. Then also the boundary condition on the observation directions (3.52) gains an intuitive physical explanation. If the observation direction $\boldsymbol{n}_1$ is chosen such, that Eq. (3.52) is not satisfied, then Eq. (3.51) equally has no solution for real values of x^-. This implies that the electron during its sojourn in the laser field, at no instant points into directions $\boldsymbol{n}_1$, which satisfy Eq. (3.52). This boundary condition on the observation directions, which was found in a quantum computation, simply states that in the relativistic regime no emission can be observed under directions into which the electron does not propagate. We stress, that in contrast to this classical analogy for the angular distribution of the electron's emission, in the frequency distribution quantum effects have to be taken into account.

3.3.2 Probability Structure in the Relativistic Regime

Inserting the stationary point expansion into the matrix element yields an asymptotic expansion of the overall transition amplitude and thus of the transition probability. We can write the stationary point expansion of the scattering matrix element as

$$S_{fi} = \sum_{l=1}^{N_{\text{s.p.}}} S_{fi}^{(l)} \tag{3.53}$$

where l is an index running over all $N_{\text{s.p.}}$ stationary points of the dynamic integrals. As we saw in the previous section the stationary points correspond to those segments of a classical trajectory, where an electron propagates and emits into the direction $\boldsymbol{n}_1$. The quantum result in the high intensity limit consequently offers a quasi-classical interpretation as a probability which is generated at all possible emission points where a classical electron emits into the direction of observation. The probability, obtained from the modulus square of Eq. (3.53) will hence contain the sum of the direct contributions from each stationary point and the respective interference terms

$$\frac{dW}{d\omega_1 d\Omega_1} \propto \sum_{l} |S_{fi}^{(l)}|^2 + \sum_{l,k} \text{Re}\left(S_{fi}^{(l)} S_{fi}^{(k)*}\right), \tag{3.54}$$

where the δ-function was already used to fix the electron's final momentum. As their origin is interferences from different possible points of emission, the terms of the double sum in Eq. (3.54) are expected to oscillate in in the final particles phase space variables such as photon energy. This is in fact the case as is seen in the following way: The contribution of every stationary point, as given in Eq. (3.36), is represented as a sum over the three integral approximations $\mathcal{I}_i^l$. Each of these integrals itself is oscillating in the final particles' phase space variables through the exponential factor $\exp\left[-\mathrm{i}S_{p_i,p_f}^{k_1}(\mathring{x}^{-,l})\right]$. These factors cancel in the direct squares, but give rise to oscillations in the interference terms of the form $\exp\left[-\left(\mathrm{i}S_{p_i,p_f}^{k_1}(\mathring{x}^{-,l}) - S_{p_i,p_f}^{k_1}(\mathring{x}^{-,k})\right)\right]$. Assuming that the dynamic integrals could be approximated by Eq. (3.36) in the first place, we tacitly supposed that the stationary points are pairwise wide enough apart, to treat their contributions separately. Please note that due to the oscillating character of $\psi_{\mathcal{A}}(x^-)$, this assumption is usually well justified. This corresponds to the requirement

$$\int_{\mathring{x}^{-,l}}^{\mathring{x}^{-,k}} \psi_A(c^-)dc^-, \int_{\mathring{x}^{-,l}}^{\mathring{x}^{-,k}} \psi_A^2(c^-)dc^- \sim 1. \tag{3.55}$$

It is then obvious that the difference scales as the exponential phases themselves in the case of large ξ

$$\left[S^{k_1}_{p_i,p_f}(\mathring{x}^{-,l}) - S^{k_1}_{p_i,p_f}(\mathring{x}^{-,k})\right] \sim S^{k_1}_{p_i,p_f}(\mathring{x}^{-,l}) \sim \xi^3. \tag{3.56}$$

We conclude that the interference terms in Eq. (3.54) are rapidly oscillating in the final photon frequency. When averaging over a relatively small frequency interval, as is done for instance in a detector with a realistic spectral resolution, these oscillations cancel, i.e. the interferences cannot be resolved. It is thus customary to neglect the contribution of the double sum in Eq. (3.54) for large ξ and to only consider the direct contributions from each stationary point separately. The total probability can then be written as

$$\frac{dW}{d\omega_1 d\Omega_1} = \mathcal{N}^2_{\text{NSCS}} \sum_l \left|\bar{u}_{p_f} M^{(l)}_{fi,2} u_{p_i}\right|^2. \tag{3.57}$$

3.4 Determining the Carrier Envelope Phase of an Intense Few-Cycle Laser Pulse

In this section we wish to sketch a possible technical application of investigating the single photon emission spectra of a relativistic electron, scattered from an ultra-intense laser pulse. It was shown that these emission spectra are sensitive to the CEP of the scattering laser pulse x_0^- [14]. Since the outlined scheme in principle is applicable for arbitrary central frequencies of the scattering laser pulse ω_L, however, it is favourable to discuss the dependence on $\eta_0 = k_L^+ x_0^-$, rather than on x_0^- itself. To highlight the importance of a CEP determination scheme for relativistically intense laser pulses, we recall the intimate connection between the generation of ultra-intense laser pulses with the generation of ultra-short pulses [16]. As was mentioned in Chap. 1, it is nowadays common practice to generate laser pulses containing only very few, even down to only one single electromagnetic field oscillations. In contrast to a monochromatic laser wave, such a laser pulse containing only a few cycles of the carrying electromagnetic wave is not fully described by its carrier frequency and its intensity. Much rather its precise shape, as e.g. modeled by Eq. (2.6), depends among others on the number of cycles the pulse contains, the precise shape of the envelope function as well as η_0. Its effect can be thought of as follows: Fixing the amplitude, frequency as well as the duration of a laser pulse, the resulting pulse is not yet fully determined, as the absolute phase of the laser pulse is not known. By virtue of the 2π periodicity of the carrier wave this absolute phase can be mapped to the CEP. If an elementary particle or atomic system now interacts with a few-cycle pulse,

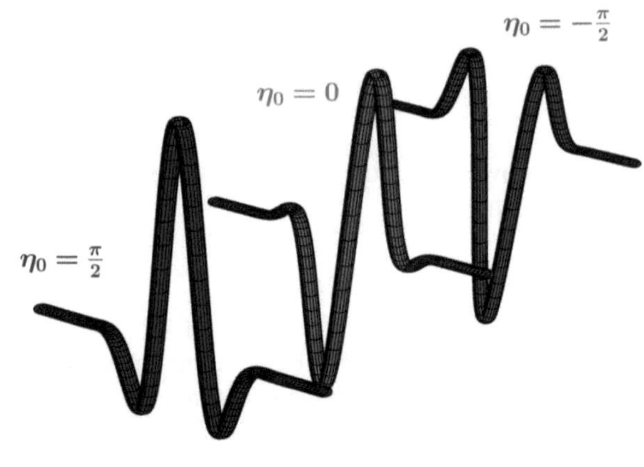

Fig. 3.8 Different pulse forms for varying CEP η_0

its dynamic behaviour will consequently depend on more quantities than the laser's frequency and its intensity. In fact, changing the carrier envelope phase of a few-cycle pulse may significantly change the pulse' s electric field shape and hence a particle's dynamics, as is shown in Fig 3.8. The strong differences between the depicted pulse forms immediately point out the importance of determining the CEP. Furthermore, control over the CEP would immediately give access to a fine control over particle dynamics. In fact, the generation of arbitrarily shaped electric field forms is one of the main goals of the PFS project [17]. It is consequently of high importance to determine and possibly manipulate the CEP of few-cycle laser pulses. Up to nowadays this task can be achieved by means of several schemes such as attosecond streaking [18], THz-spectroscopy [19] or a stereo above-threshold ionization (ATI) measurement [20, 21]. We wish to shortly sketch the concepts of the latter, as depicted in. Fig 3.9.[1] We also point out its intrinsic deficiency, not to be capable of analyzing highly intense pulses. For a stereo ATI measurement the laser pulse, whose CEP is to be determined, is brought into collision with a beam of atoms (e.g. Xenon as in the original work [20]). The laser pulse will then ionize the atoms and accelerate the freed electrons towards the direction of its instantaneous electric field. After passing the interaction region, the laser pulse is detected by a photo diode (PD), while the photoelectrons are guided through a time-of-flight spectrometer onto two microchannel plates at the respective faces of the interaction region. In this way a stereo ATI measurement can compare the abundance of electrons accelerated to the respective directions of electric field orientation and thus determine the relative abundance of "positive" and "negative" electric field spikes (here "positive" and "negative" refers to opposing directions with respect to an arbitrarily chosen axes along the laser's polarization direction). From this signal one can then reconstruct the pulse's CEP. The proposed scheme, however, suffers a serious deficiency. If the laser pulse in question is intense enough to saturate ionization of the atoms in the beam before it has fully passed the interaction region, the stereo ATI method does not give a signal from the whole pulse.

[1] Reprinted figure with permission from [20]. Copyright (2003) by the American Physical Society.

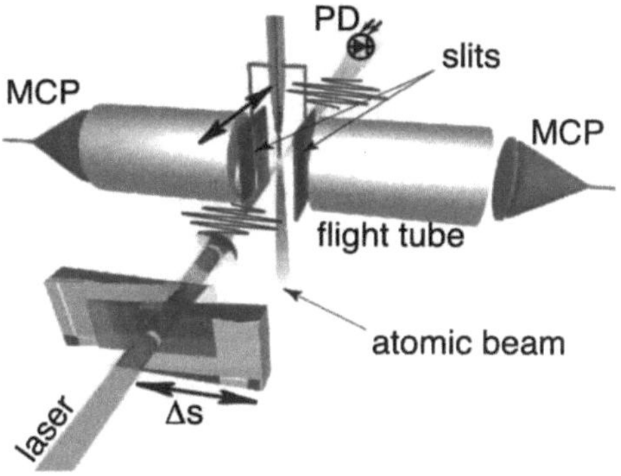

Fig. 3.9 Principle of a stereo ATI measurement (*picture* from [20]). *PD* Photo diode, *MCP* Microchannel plates

This scheme, however, relies crucially on the emission of electrons over the complete duration of the laser pulse, as the detected electron yield is an integrated quantity and thus the photoelectron yield must be continuous over the whole pulse duration. Since now the ionization energies of most atomic species, that are technically usable in laboratory experiments, lie in the eV-keV regime, a laser only needs to provide comparatively small energies to a single electron to ionize it from its parent atom. A laser pulse in the regime $\xi \sim 1$, on the other hand, provides an energy on the order of the electron's rest mass within one cycle of the electric field, whence we immediately conclude that conventional CEP determination schemes are limited to the realm $\xi \ll 1$. Bearing in mind that one of the main motivations to temporally compress laser pulses as tightly as possible, is to obtain highest laser intensities, the lacking possibility of fully characterizing such ultra intense laser pulses is clearly dissatisfactory. Employing the dependence of the angular radiation distribution of NSCS spectra could close this gap [14].

The proposed scheme can be explained by some simple principle considerations: The radiation emitted by an ultra-relativistic electron with instantaneous energy ε is confined to an emission cone of aperture $m/\varepsilon \ll 1$ along its instantaneous velocity (see Sect. 2.1). One can then picture that the classical emission of an electron maps its trajectory like a torchlight. The trajectory in turn, depends on the scattering pulse's field structure and in particular on its CEP. Thus the angular distribution can serve to map the CEP. Formally the method relies on the stationary phase analysis outlined in Sect. 3.3. There we found that a real stationary point can be found only, if the shape function fulfills condition (3.52), confining the possible values of the ratio $-\alpha_V/2\beta_V$. For definiteness, from here on we consider the process in the coordinate system specified in Fig. 2.1. In the relativistic regime $\varepsilon_i, m\xi \gg m$, Eq. (3.52) then turns into

$$\frac{(m\xi)}{2\varepsilon_i}\psi_{\mathcal{A}}^{\min} \leq \cos(\varphi_1)\cot\left(\frac{\vartheta_1}{2}\right) \leq \frac{(m\xi)}{2\varepsilon_i}\psi_{\mathcal{A}}^{\max}. \tag{3.58}$$

This expression depends on the electron's initial energy and the laser intensity only via the simple ratio $m\xi/\varepsilon$. Considering an experimental situation where the electron's initial energy and the laser intensity are fixed, we can consider the photon's emission angles ϑ_1 and φ_1 as free parameters in the above equation. Since the absence of any real stationary points leads to an exponential damping of the dynamic integrals f_i, we predict that there is no emission in observation regions (ϑ_1, φ_1), where Eq. (3.58) is not fulfilled. For this equation to be satisfied, because of $\psi_\mathcal{A} \sim 1$ it must hold $\cot(\vartheta_1) \lesssim {}^{m\xi}/_{\varepsilon_i}$. (recall $|\cos(\varphi_1)| \leq 1$). For the azimuthal emission angle φ_1 we deduce a scaling condition from Eq. (3.26). For $\varepsilon \gg m\xi$ it needs to scale as $\varphi_1 \lesssim {}^{m}/_{\varepsilon_i}$, whereas for $\varepsilon_i \ll m\xi$ it must satisfy $\varphi_1 \lesssim {}^{1}/_{\xi}$, since elsewise the imaginary contribution κ is not strongly suppressed and the dynamic integrals are damped. These angle ranges are in agreement with the classical considerations presented at the beginning of Sect. 3.3. Since now the precise values of $\psi_\mathcal{A}^{\min}$ and $\psi_\mathcal{A}^{\max}$ depend on the CEP η_0, Eq. (3.32) can be written as

$$\psi_\mathcal{A}^{\min}(\eta_0) \leq -\frac{\alpha}{2\beta} \leq \psi_\mathcal{A}^{\max}(\eta_0). \tag{3.59}$$

where η_0 is considered as a free parameter. Since the electron's emission is confined to the plane of polarization of the laser (recall the scaling laws of φ_1), we can distinguish the direction of observation by the single parameter ϑ_1 by mapping $\varphi_1 = 0$ to $\vartheta_1 > \pi$ and $\varphi_1 = \pi$ to $\vartheta_1 < \pi$, respectively. Equation (3.58) then turns into a boundary condition for the polar emission angle in the relativistic regime

$$\pi - \operatorname{arccot}\left(-\frac{m\xi}{2\varepsilon_i}\psi_\mathcal{A}^{\min}(\eta_0)\right) \geq \frac{\vartheta_1}{2} \geq \operatorname{arccot}\left(\frac{m\xi}{2\varepsilon_i}\psi_\mathcal{A}^{\max}(\eta_0)\right). \tag{3.60}$$

Please note that the additional -1 on the left side of Eq. (3.60) is introduced by $\cos(\varphi = \pi)$ and the additional term π performs the desired mapping to $\vartheta_1 > \pi$. We furthermore note that due to the normalization $|\psi_\mathcal{A}(\eta)| < 1$ the emission angle ϑ_1, independently of the CEP η_0, is always confined to an emission cone

$$|\pi - \vartheta_1| \leq \vartheta_0, \tag{3.61}$$

where the maximum angle is given by $\vartheta_0 = \pi - 2\operatorname{arccot}({}^{m\xi}/_{2\varepsilon_i})$. In the important case $\varepsilon_i \gg m\xi$ it is approximately given by $\vartheta_0 \approx {}^{m\xi}/_{\varepsilon_i}$. We thus find that this important ratio can be viewed as the invariant boundary angle of the emission cone of a highly relativistic electron scattered from an intense laser pulse, irrespective of the scattering pulse's specific shape.

To give explicit numerical values we consider a plane wave laser field of the temporal shape Eq. (2.6) with $n_C = 2$. For an optical laser of frequency $\omega_L = 1.55$ eV, this corresponds to a pulse duration $\tau_L \approx 5$ fs. As usual we consider the scattering in the reference frame shown in Fig. 2.1. Furthermore we imagine that it is favourable to employ laser accelerated electron beams in the outlined experimental setup, since the spatial extensions of such beams approximately match the extent of the accelerating laser pulse, whence a considerable overlap is envisaged. Furthermore it has been demonstrated that laser acceleration schemes become increasingly compatible

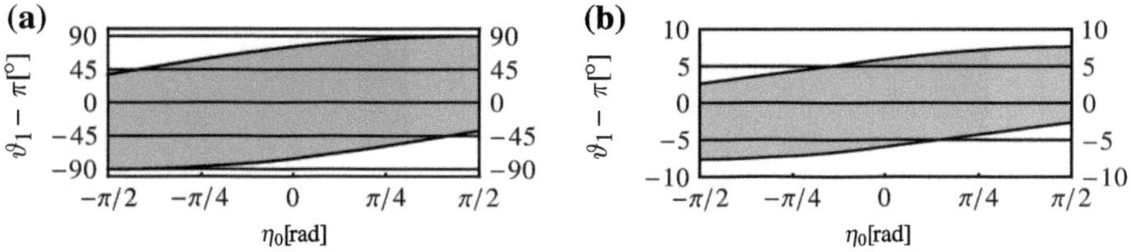

Fig. 3.10 Change of angular emission range with changing CEP η_0 for the indicated ratios of $\varepsilon_i/m\xi$. The range into which an electron is predicted to radiate according to Eq. (3.58) is *gray*. **a** $\varepsilon_i = m\xi/2$. **b** $\varepsilon_i = 7.5m\xi$

with conventional accelerators, concerning the achievable energies, particle densities as well as beam focusing [22–26]. Finally due to the far simpler experimental implementation of laser acceleration schemes in laser laboratories, as compared to the combination of high-power laser facilities with state-of-the-art particle accelerators, in this work we are to going consider experimental specifications of laser generated electron beams. In Fig. 3.10a we plot relation Eq. (3.60) for $m\xi = 2\varepsilon_i$. In this figure the gray shaded area gives the range where emission is expected, in dependence of the CEP η_0. We observe a clear dependence of the emitted radiation's angular distribution on the CEP. In turn, measuring the cutoff angles of the angular range, into which an electron emits radiation, it is in principle simple to read off the scattering laser pulse's CEP. In Fig. 3.10b we plot relation Eq. (3.60) for a parameter ratio of $\varepsilon_i = 7.5\,m\xi$. In this figure we find a considerably smaller angle range where radiation may be emitted to as compared to Fig. 3.10a. This can be expected however, since, due to the increased ratio $\varepsilon_i/m\xi$, the electron is less strongly deflected in the laser pulse and emits into an emission cone confined narrower around its initial propagation direction at $\vartheta_1 = \pi$.

To demonstrate now the accuracy and potential of the outlined CEP determination scheme, we compare the analytical approximation Eq. (3.60) to exact numerical simulations of Eq. (3.21), which are obtained by the quadrature scheme outlined in Chap. 5. We will always consider a laser pulse modeled by Eq. (2.6) with $n_C = 2$. First we show an example chosen to match the case of Fig. 3.10a. Hereby we demonstrate the proposed method's applicability to the laser system PFS, already under construction at Garching [17]. We consider a laser intensity of $I = 2 \times 10^{22}\,\mathrm{W/cm^2}$ (corresponding to an intensity parameter of $\xi \approx 10^2$). The initial electron energy is set to 25 MeV such that $m\xi \approx 2\varepsilon_i$ as was used in Fig. 3.10a. We point out that in this case $\chi \approx 2 \times 10^{-2}$, therefore quantum effects are negligible. In Fig. 3.11a, b we show two energy spectra calculated from Eq. (3.21) for the two different CEPs $\eta_0 = -\pi/10$ (Fig. 3.11a) and $\eta_0 = -\pi/5$ (Fig. 3.11b). In both figures the colored regions give the exact energy emission spectrum numerically obtained from Eq. (3.21) and the white horizontal lines are the CEP-dependent cutoff angles of the emission cone, as obtained from Eq. (3.60). The solid black line appended to the left of both spectra gives the total emission into that specific angle, integrated over all frequencies in arbitrary units (the scale tends to larger values to the left). By comparing the posi-

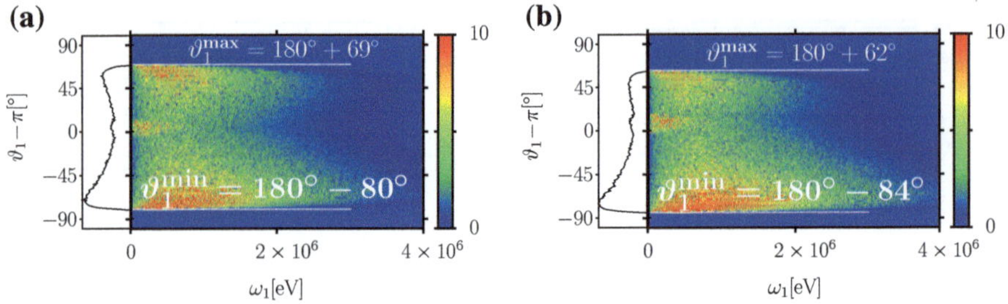

Fig. 3.11 Energy emission spectra $dE/d\Omega_1 d\omega_1$ in sr^{-1} for $\omega = 1.55\,\text{eV}$, $\tau = 5.4\,\text{fs}$, $I = 2 \times 10^{22}\,\text{W/cm}^2$, $\varepsilon_i = 25\,\text{MeV}$. The *horizontal white lines* are placed at the analytically predicted cutoff angles ϑ_1^{min}, ϑ_1^{max} corresponding to Eq. (3.60). **a** Quantum simulation for $\eta_0 = -\pi/10$. **b** Quantum simulation for $\eta_0 = -\pi/5$

tions at which this integrated angular spectrum goes to zero rather rapidly to the analytical prediction of Eq. (3.60), we can additionally check the predictive power of that equation. In fact, an excellent agreement between the analytical prediction and the numerical simulation is apparent. The method's sensitivity can be inferred from the change of the cutoff angles, apparent in Fig. 3.11b. The CEP employed for obtaining this spectrum is changed with respect to Eq. (3.60) by $\Delta\eta_0 = \pi/10$ to $\eta_0 = -\pi/5$. In Fig. 3.11 we again observe the agreement between Eq. (3.60) and the numerical simulation. We furthermore observe that the upper (lower) cutoff angle is changed by 7° (4°). Since this should in principle be detectable in experiment, we claim a CEP sensitivity of at least $\Delta\eta_0 = \pi/10$. This is comparable to the first reported sensitivity of the conventional stereo ATI scheme, when it was introduced [20]. Nowadays, however, conventional CEP determination techniques can reach up to sensitivities of $\pi/300$ [21].

To demonstrate how quantum effects may affect the emission spectra, we consider a changed set of parameters. The scenario we have in mind, is the interaction of an XUV laser pulse of central frequency $\omega_L = 50\,\text{eV}$ and intensity $I_L = 10^{24}\,\text{W/cm}^2$ ($\xi \approx 20$) [27] with an electron, initially having an energy of $\varepsilon_i = 75\,\text{MeV}$. In Fig. 3.12a we show the energy emission spectrum for the specified parameters and a CEP of $\eta_0 = 0$, while in Fig. 3.12b we change the CEP to $\eta_0 = \pi/4$. In both figures we indicate the maximal quantum mechanically allowed energy, the emitted photon can have according to Eq. (3.16), by a black line on the right side of the spectra. We find that this cutoff energy is well approached in both spectra. In addition, for the chosen parameters we obtain $\chi \approx 0.6$, whence we conclude that quantum effects in fact play a role. Comparing a classical emission spectrum for the same experimental parameters as in Fig. 3.12a and a CEP $\eta_0 = 0$ to the quantum result, confirms this assertion. The result is shown in Fig. 3.12c with the integrated angular spectrum at the left taken from the quantum simulation. Comparing now Fig. 3.12a, c we conclude that, while the cutoff-angles are well predicted by the classical simulation, the cutoff frequency is largely exceeded. The former observation is a further confirmation that the picture of a classical electron trajectory, from which the emission's angular distribution

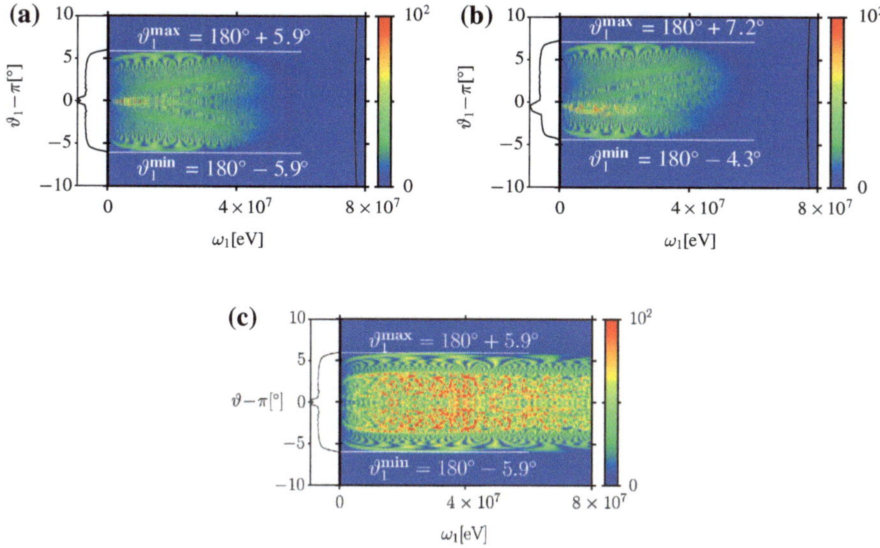

Fig. 3.12 Energy emission spectra $dE/d\Omega_1 d\omega_1$ in sr^{-1} for $\omega = 50\,eV$, $I = 10^{24}\,W/cm^2$, $\varepsilon_i = 75\,MeV$. The *horizontal white lines* are placed at the analytically predicted cutoff angles ϑ_1^{max} corresponding to Eq. (3.60) and the *vertical black line* on the *right* gives the quantum mechanical cutoff energy for the emitted photon ω_1^{Max} according to Eq. (3.16). **a** Quantum simulation for $\eta_0 = 0$. **b** Quantum simulation for $\eta_0 = \pi/4$. **c** Classical simulation for $\eta_0 = 0$

can be inferred, inside an ultra-intense laser field remains largely valid. The latter effect, however, proves that photon emission from an electron has to be treated as a quantum effect, since by calculating the classical energy emission spectrum by means of the Lienard-Wiechert potentials, one assumes the particle's trajectory to be unaffected by the photon emission. The electron can then emit arbitrarily high photon energies, what is forbidden by energy-momentum conservation in QED. Coming back now to Fig. 3.12a, b we observe that the proposed experimental scheme provides a reliable CEP determination scheme also in the case that photon recoil is considerable. However, since the overall emission cone is less wide opened, as compared to the scenarios studied in Fig. 3.11, a fixed change $\Delta\eta_0$ in the CEP results in a comparatively smaller change of the cutoff angles. Nevertheless in Fig. 3.12a, b for a CEP change $\Delta\eta_0 = \pi/4$ we observe a change of the upper (lower) cutoff angle of 1.3° (1.6°). Though more difficult to detect this change should still be accessible to experimental validation although the proposed method is less accurate as compared to the previous case.

Finally we wish to comment on a possible fallacy concerning the proper choice of the ratio $m\xi/\varepsilon_i$. Since with a larger opening angle of the emission cone there obviously comes an increased change of the cutoff angles for a fixed CEP change $\Delta\eta_0$ one might think, that for a given laser intensity $I(\xi)$ it would be most advantageous, to choose an electron with significantly smaller energy $\varepsilon_i \ll m\xi$ scattering from it. This, however, is not the case since in such a setup, the electrons would be almost immediately

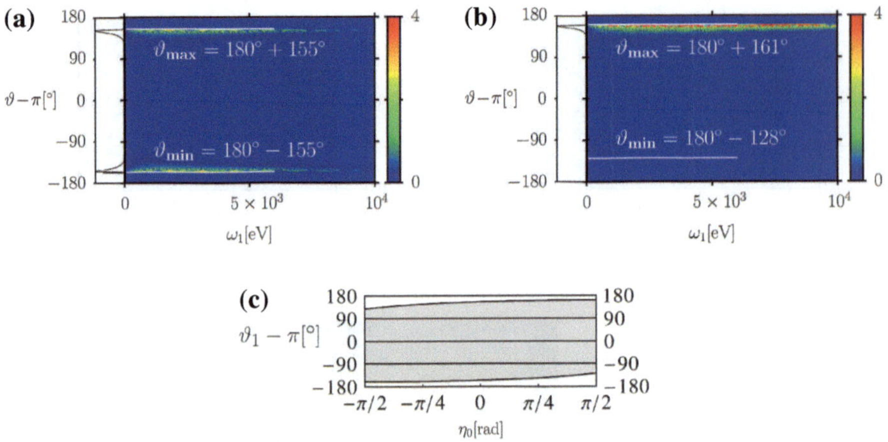

Fig. 3.13 Energy emission spectra $dE/d\Omega_1 d\omega_1$ in sr^{-1} for $\omega = 1.55\,eV$, $\tau = 5.4\,fs$, $I = 10^{21}\,W/cm^2$, $\varepsilon_i = 2\,MeV$. The *horizontal white lines* are placed at the analytically predicted cutoff angles ϑ_1^{min}, ϑ_1^{max} corresponding to Eq. (3.60) and shown in Fig. 3.13c. **a** Quantum simulation for $\eta_0 = 0$. **b** Quantum simulation for $\eta_0 = \pi/2$. **c** Angular emission range for $m\xi \gg \varepsilon_i$ according to Eq. (3.60)

scattered back from the strong laser wave and emit almost exclusively into a narrow cone around the laser's propagation direction with opening angle $\varepsilon_i/m\xi \ll 1$. Changes in this cutoff angles would again be difficult to observe. To demonstrate this effect in Fig. 3.13 we show energy emission spectra from an electron with an initial energy of $\varepsilon_i = 2\,MeV$, scattered from an optical ($\omega_L = 1.55\,eV$) laser pulse with an intensity of $I = 10^{21}\,W/cm^2$, corresponding to a nonlinearity parameter of $\xi \approx 20$). These choices correspond to an only mildly relativistic initial electron velocity of $|\boldsymbol{\beta}_i| \approx 0.86$ and a ratio $\varepsilon_i/m\xi \approx 10^{-1}$. Indeed we observe, that the electron radiates almost exclusively into regions close to the cutoff angles of the emission cone. It is not possible, however, to use the cutoff angles of an emission cone around the observation direction $\vartheta_1 = 0°$ in a similar way as described before, since the cutoff angles in between the two white lines in Fig. 3.13a, b are not sharply defined, but rather washed-out (compare the integrated angular emission distribution left of the color-coded spectra). Furthermore in Fig. 3.13c we observe that the most prominent change in the cutoff angles is found in ϑ_{min} (ϑ_{max}) for $\eta_0 > 0$ ($\eta_0 < 0$). This observation is also confirmed in Fig. 3.13a, b. In this direction, however, as can be seen in Fig. 3.13b virtually no emission is detectable. For an even smaller ration $\varepsilon_i/m\xi$ this behaviour would even worsen. It is consequently most sensible to choose ε_i and $m\xi$ of the same order, to perform a CEP measurement.

3.4.1 Experimental Limitations

To consider a realistic experimental scenario we cannot resort to a single electron scattered from a plane wave. Much rather in an experiment an electron beam is brought to collision with a spatially focused laser. This section is devoted to the discussion of possible experimental limitations of the proposed CEP determination scheme. First we discuss a particular experimental challenge. In fact, for obtaining values of the nonlinearity parameter $\xi \gtrsim 1$ one has to consider ultra-high intensity laser facilities. In all of these, tremendous laser intensities are obtained by tight temporal and spatial focusing of the laser pulse. Whereas temporal focusing is exactly taken into account in our calculations, spatial focusing is not included. This approximation of the laser pulse as a plane wave, however, could possibly be problematic, since it completely neglects possible changes in the laser pulse's peak intensity parameter ξ during the interaction. To understand this issue we recall from Sect. 2.1.3 that a laser focus is confined perpendicularly to the laser's propagation direction by its beam waist w_0. In its propagation direction it stretches over its Rayleigh length $l_R = 2\pi w_0^2/\lambda_L$. For a laser pulse, focussed to its diffraction limit $w_0 \approx \lambda_L$, this corresponds to a longitudinal dimension of $l_R \sim \lambda_L$. The typical stretch of a laser accelerated electron bunch, on the other hand, is on the order of half a plasma wavelength $\Delta x_{e^-} \sim \lambda_P/2 = \pi/\omega_p = \sqrt{m\pi/4n_e e^2}$, where ω_p is the plasma frequency and n_e the electron density in the plasma. For typical densities $n_e \sim 10^{18}\,\text{cm}^{-3}$ this translates to stretches on the order of $\Delta x_{e^-} \sim 10\,\mu\text{m}$, exceeding the Rayleigh range by more than an order of magnitude. The laser pulse thus will pass the focal spot on a significantly shorter time scale than the electron bunch. Since outside of the focal spot, however, the laser's intensity strongly differs from its peak value, the electrons will experience a varying laser peak intensity. The resulting variation of the parameter ξ is not accounted for in Eq. (3.60). Consequently one might wonder, how the proposed scheme is supposed to determine a reasonable result for the CEP. To correctly include spatial focusing in our quantum approach, according to Sect. 2.2, one would have to employ a solution of the Dirac equation in the presence of a spatially focussed laser field. A perturbative series of such a solution is presented in Appendix C, but a closed expression is not known up to date. To nevertheless investigate the impact of spatial focusing on the applicability of the proposed scheme, we will take it into account by replacing the results in the classical regime by actual classical computations, where spatial focusing can be incorporated. To this end we first compare the quantum simulation for $\chi \ll 1$ from Fig. 3.11 to a classical simulation for a wide laser focus. In Fig. 3.14 we show a classical energy emission spectrum from a single electron with initial energy $\varepsilon_i = 25$ MeV, and colliding head-on with a laser pulse modeled by a Gaussian beam focus of Eq. (2.6), focussed to a spot size of $w_0 = 200\,\mu\text{m} > 10^2\lambda_L$ with a peak intensity $I = 2 \times 10^{22}\,\text{W/cm}^2$. Furthermore we consider the electron to be initially propagating on the focal axis of the laser pulse. The focusing parameter evaluates to $s < 10^{-2}$, whence we deduce that the plane wave approximation is in good agreement with the spectra shown in Fig. 3.11 as is expected. We note that to achieve the desired laser intensity at the

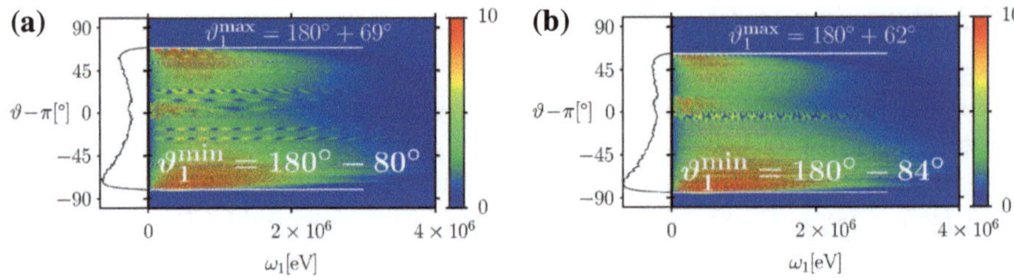

Fig. 3.14 Classical simulation of the energy emission spectra $dE/d\Omega_1 d\omega_1$ in sr^{-1} for $\omega = 1.55$ eV, $\tau = 5.4$ fs, $I = 2 \times 10^{22}$ W/cm^2, realized by a Gaussian beam focused to $w_0 = 200\,\mu$m and $\varepsilon_i = 25$ MeV. The horizontal white lines are placed at the analytically predicted cutoff angles $\vartheta_1^{\min}$, $\vartheta_1^{\max}$ corresponding to Eq. (3.60). **a** *CEP* $\eta_0 = -\frac{\pi}{10}$. **b** *CEP* $\eta_0 = -\frac{\pi}{5}$

considered focal spot size, one would have to realize a laser system with an overall energy of $E_L \approx 120$ kJ, corresponding to a laser power of $P_L \approx 20$ EW, delivered within 5 fs. Although this is well out of experimental reach, we proceed considering such a hypothetical laser system, to analytically confirm the simulations. And as expected we observe a fine agreement between Fig. 3.11 and 3.14. The classical simulations thus can be employed to investigate the impact of spatial focusing of the laser pulse in the classical regime, by departing from the considered parameters to experimentally feasible ones. To investigate this effect for a most realistic setting, in Fig. 3.15 we show the emission spectra of an electron bunch, featuring a Gaussian energy distribution with a mean energy of $\varepsilon_i = 25$ MeV an a 1 % spread [22–26], and a spatial extent parallel (perpendicular) to its propagation direction of 8 μm (5 μm). The Gaussian distribution function of the electron bunch is normalized to a total number density equal to unity. This bunch is assumed to collide head-on with a Gaussian focussed laser pulse equal to that studied in Fig. 3.14, but focussed to a spot size of $w_0 = 2\,\mu$m. To describe the focusing of a few-cycle laser pulse to such small focal spots, we employ the discussion of Sect. 2.1.3. The power required, to generate a laser intensity of $I_L = 2 \times 10^{22}$ W/cm^2 is consequently reduced to $P_L \approx 2$ PW, as is well achievable at present day laser facilities, such as the PFS in Garching [17]. On the left of the spectra we again show the total emission into the specified polar angle in arbitrary units. We observe significant changes in comparison to the case of a big laser focus, as shown in Fig. 3.14. Firstly we note that due to interferences from the many electrons in the bunch, the spiky structure of the single electron spectra of Figs. 3.11 and 3.14 is largely washed out. More important we note a strong decrease in the overall emitted energy. We draw special attention to the reduced scales of the color-code and ω_1-axis. It is however, reasonable to expect less energy to be emitted from the interaction of an electron bunch scattered from a tightly focussed laser pulse, since the scattered electrons experience the highest possible field strengths in the laser pulse only in a comparatively small spatial region. Despite the strong differences in emitted energy we also observe in Fig. 3.15, that the cutoff-angles of the emission region are well predicted by the plane wave prediction Eq. (3.60), also for a realistically reduced focal spot size. We recall that

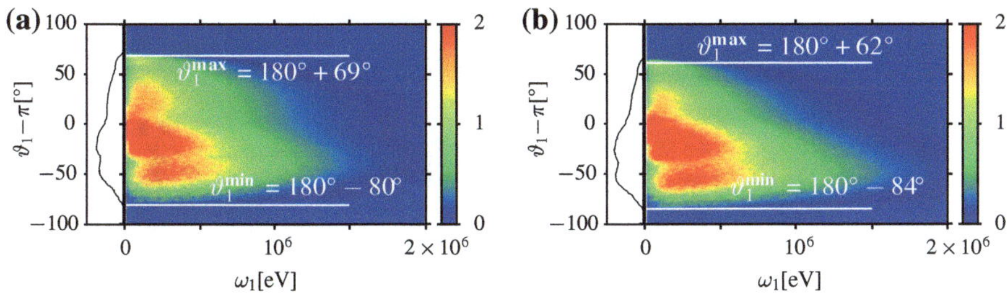

Fig. 3.15 Classical simulation of the energy emission spectra $dE/d\Omega_1 d\omega_1$ in sr^{-1} for $\omega = 1.55$ eV, $\tau = 5.4$ fs, $I = 2 \times 10^{22}$ W/cm^2, realized by a Gaussian beam focused to $w_0 = 2\,\mu$m scattering an Gaussian electron bunch with a mean energy of $\varepsilon_i = 25$ MeV. The *horizontal white lines* are placed at cutoff angles $\vartheta_1^{\min}$, $\vartheta_1^{\max}$ analytically predicted by the plane wave analysis given in Eq. (3.60). **a** *CEP* $\eta_0 = -\frac{\pi}{10}$. **b** *CEP* $\eta_0 = -\frac{\pi}{5}$

this equation is derived from the exponential phase of the scattering matrix element. In Appendix C we find that, to describe the effect of the laser focusing on this phase in lowest order of the focusing parameter s, one simply has to replace the plane wave shape function $\psi_{\mathcal{A}}(x^-)$ with its focused counterpart $\psi_{\mathcal{A}}(x^-)\Psi_L(\boldsymbol{x})$, according to Eq. (2.45). This procedure results in a slight change of the predicted cutoff angles and further improves the agreement with the numerical simulation [14]. The good agreement between the cutoff angles of the plane wave and the focused beam analyses is explainable looking back at Eq. (3.60): Decreasing the value of ξ decreases the minimum and maximum values the right and left hand sides of that inequality can take. This corresponds to a contraction of the emission cone. Thus the cutoff angles of the overall emission cone are formed at the regions of maximal laser intensity. The proposed scheme thus effectively only probes the focal region of the scattering laser field. Consequently the CEP determination is expected to work as long as the scattering laser's focussed intensity is sufficiently well known and the electron density in the bunch sufficiently high to probe all portions of the interaction region.

Further effects, which are not captured in Eq. (3.60) and which we wish to discuss briefly are:

1. multi-particle effects in the electron bunch, as Coulomb repulsion or coherence in the emission of separate electrons
2. finite size effects from the extent of the focal spot
3. uncertainty in laser intensity I_L and mean electron energy ε_i
4. choice of the envelope function model.

The first disturbing effect is most easily estimated. Electron beams from laser acceleration schemes are reported to feature divergences of $\Delta\epsilon \approx 5$ mrad, and Coulomb repulsion is included in this estimate. Considering the electrons to be generated in a focal spot according to Fig. 3.16, we can estimate that the electron's initial propagation direction will deviate from the negative z-axis by an angle $\Delta\epsilon/2$ at most. Approximating $\Delta\epsilon/2 \gg m/\varepsilon_i$, which is a good approximation in the relativistic regime,

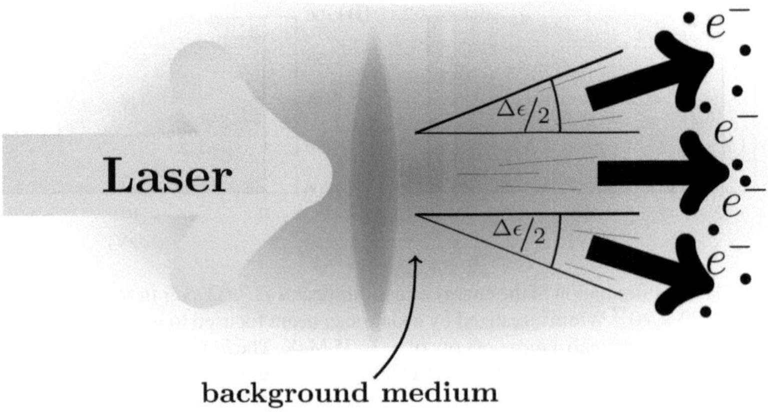

Fig. 3.16 Schematic electron bunch acceleration

this deviation from a head on collision coincides with the uncertainty in the cutoff angles. In our analysis of Eq. (3.60) we found $\Delta\vartheta_1 \approx 1 - 10°$ which largely exceeds typical values of $\Delta\epsilon/2$. For an electron bunch of sufficiently small divergence electron-electron interaction is consequently negligible in our analysis. To give an estimate for the impact of interferences between the emissions from separate electrons we note that a typical electron density in a laser accelerated bunch is on the order of $10^{18}\,\text{cm}^{-3}$. This value corresponds to an average spacing between the single electrons of 1 μm in every space dimension. In the previous sections we saw that the dominant contribution to the emitted radiation is in the frequency range $\omega_1 \sim \omega_L\xi^3 \gg \omega_L$. For the interaction with optical lasers this radiation exhibits a coherence interval on the order of $\lambda^{\text{coh}} \sim \omega_1^{-1} \ll 1\,\mu\text{m}$. The radiation emitted by two differing electrons thus is incoherent. The joined spectra emitted by all electrons in the bunch can consequently be approximated by an integral over differing initial electron parameters entering Eq. (3.21).

Secondly one has to estimate the uncertainty in the angle determination resulting from the spatial extent of the interaction region. Since typical dimensions of a laser accelerated electron bunch largely exceed achievable laser focus dimensions, we consider the dimension of the interaction region to be given by the extent of the laser's focal spot. The transversal extent of this spot translates to an error in the determination of the cutoff angles. From geometrical consideration (see Fig. 3.17), we deduce that for a transversal extent of the interaction region d and a detector placed at a distance r, the corresponding uncertainty in the cutoff angles $\vartheta_1^{\text{c.o.}} \in \{\vartheta_1^{\max}, \vartheta_1^{\min}\}$ is given by $\Delta^{\perp}\vartheta_1^{\text{c.o.}} \approx d/r\,\left|\cos(\vartheta_1^{\text{c.o.}})\right|$. This error depends on the distance to the detector as r^{-1}, which reflects the fact that from great distances an extent interaction region can be approximated by a single point. In fact for all experimental scenarios we can consider, it holds $r \gg d$ and the estimated error is negligible.

The experimental uncertainties in the laser intensity and the initial electron energies, which translate to uncertainties in the parameter ξ and ε_i, can constitute another

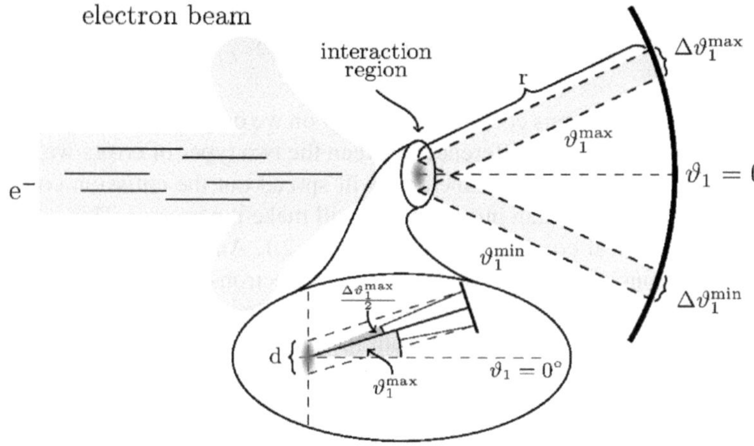

Fig. 3.17 Schematic error in the cutoff angles due to interaction regions transversal extent

limitation in the application of the proposed CEP determination scheme. Nowadays the intensity of an optical laser pulse can be determined to a relative accuracy of $\Delta I/I \approx 0.1$ [28]. Since the laser intensity scales quadratically with the parameter ξ this translates to an uncertainty $\Delta\xi/\xi \approx 5 \times 10^{-2}$. Electron energies in laser accelerated bunches, on the other hand, are reported to feature a spread of only $\Delta\varepsilon/\varepsilon \approx \times 10^{-2}$ [25, 26]. To investigate the influence of these errors on the cutoff angles $\vartheta_1^{c.o.}$, we take the derivative in Eq. (3.60) with respect to the respective parameter. The results are

$$\frac{d\vartheta_1^{c.o.}}{d\xi} = 2\frac{1}{1+\left(\frac{m\xi\psi_{\mathcal{A}}^{c.o.}}{\cos(\varphi_1)\varepsilon_i(1+\beta_i)}\right)^2}\frac{m\psi_{\mathcal{A}}^{c.o.}}{\cos(\varphi_1)\varepsilon_i(1+\beta_i)} \tag{3.62a}$$

$$\frac{d\vartheta_1^{c.o.}}{d\varepsilon_i} = -2\frac{1}{1+\left(\frac{m\xi\psi_{\mathcal{A}}^{c.o.}}{\cos(\varphi_1)\varepsilon_i(1+\beta_i)}\right)^2}\frac{m\xi\psi_{\mathcal{A}}^{c.o.}}{\cos(\varphi_1)\varepsilon_i^2(1+\beta_i)}, \tag{3.62b}$$

where $\psi_{\mathcal{A}}^{c.o.} \in \{\psi_{\mathcal{A}}^{min}, \psi_{\mathcal{A}}^{max}\}$ is connected to $\vartheta_1^{c.o.}$ via Eq. (3.58). The error dependency on an uncertainty in the parameters ξ and ε_i in the relativistic regime and observed in the coordinate system from Fig. 2.1 is then expressible as

$$\begin{aligned}\Delta^{\xi}\vartheta_1^{c.o.} &= \frac{d\vartheta_1^{c.o.}}{d\xi}\Delta\xi \\ &\approx \frac{4m\varepsilon_i\xi\psi_{\mathcal{A}}^{c.o.}}{4\varepsilon_i^2 + m^2\xi^2\psi_{\mathcal{A}}^{c.o.2}}\frac{\Delta\xi}{\xi} \end{aligned} \tag{3.63a}$$

$$\Delta^{\varepsilon_i}\vartheta_1^{c.o.} = \frac{d\vartheta_1^{c.o.}}{d\varepsilon_i}\Delta\varepsilon_i$$

$$\approx -\frac{4\xi m \varepsilon_i \psi_{\mathcal{A}}^{\text{c.o.}}}{4\varepsilon_i^2 + m^2\xi^2\psi_{\mathcal{A}}^{\text{c.o.}2}} \frac{\Delta\varepsilon_i}{\varepsilon_i}. \tag{3.63b}$$

Since the two expressions coincide, from now on we only discuss $\Delta\vartheta_1^{\text{c.o.}}$ referring to both possible errors. As a difference between the two types of errors we note that increasing the laser intensity parameter ξ will spread out the emission cone while increasing the electrons' mean initial energy will make it narrower. This is in agreement with the classical considerations from Sect. 2.1. An increased laser intensity renders the electrons' trajectories steeper whereas electrons with larger initial energy ε_i pass through the interaction region with a less prominent change of their propagation directions. Inserting the numbers from the previously discussed examples and setting $\psi_{\mathcal{A}}^{\text{c.o.}} = 1$, we obtain as an upper limit for the angular uncertainty in the classical scenario, where an optical laser was considered $\Delta\vartheta_1 \approx 2.8°$. In an analogous way we obtain for the quantum scenario an uncertainty in the cutoff angles arising from the uncertainty in the initial electron energies of approximately $\Delta^{\varepsilon_i}\vartheta_1^{\max} \approx 0.4°$. The uncertainty $\Delta^{\xi}\vartheta_1^{\max}$ for the quantum scenario, however, cannot be estimated in this way. In this scenario we considered the radiation from an XUV laser with a central frequency of $\omega_L = 50\,\text{eV}$. For such a device the intensity determination is more complicated and less accurate than for an optical laser system. To estimate the influence of uncertainties in the laser intensity, we turn the argument around to obtain the maximally tolerable uncertainty in the intensity measurement, that will not obscure the observed signal in the angular distribution. From Fig. 3.12 we read off that in order to achieve a theoretical accuracy of $\pi/4$ in the CEP, we need to know the parameter ξ to an accuracy that ensures $\Delta^{\xi}\vartheta_1^{\max} \lesssim 1°$. To invert Eq. (3.63a) most easily we observe that for the parameters we chose in the quantum scenario it holds $\varepsilon_i^2 = (7.5m\xi)^2 \gg m^2\xi^2$. We thus find

$$\frac{\Delta\xi}{\xi} \approx \frac{\Delta^{\xi}\vartheta_1^{\max}}{7.5}. \tag{3.64}$$

If we do not want to conceal an angular change of $\Delta\vartheta_{\min} = 1.6°$, as is apparent in Fig. 3.12 we can accept a relative error of $\Delta\xi/\xi \approx 0.2$. This corresponds to an intensity which is only known with an uncertainty of 40 % without obscuring the predicted CEP effect.

We next show that the specific shape function $\psi_{\mathcal{A}}(\eta)$, chosen for modeling the laser pulse, has no influence on the result. To this end we compare our standard choice of Eq. (2.6) to two different types of alternative envelopes, namely first a Gaussian and second a hyperbolic secans as envelope functions

$$\psi_{\mathcal{A}}^{\text{Gauss}} = \sin(\eta + \eta_0)\,\mathrm{e}^{-\frac{(\eta-2\pi n)^2}{a}} \tag{3.65a}$$

$$\psi_{\mathcal{A}}^{\text{sech}} = \sin(\eta + \eta_0)\,\text{sech}\left[\frac{(\eta - 2\pi n)}{a}\right]. \tag{3.65b}$$

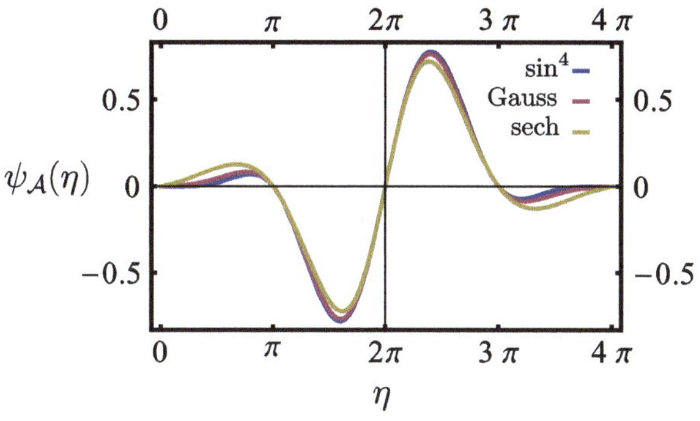

Fig. 3.18 Two-cycle laser pulses for three different envelope functions

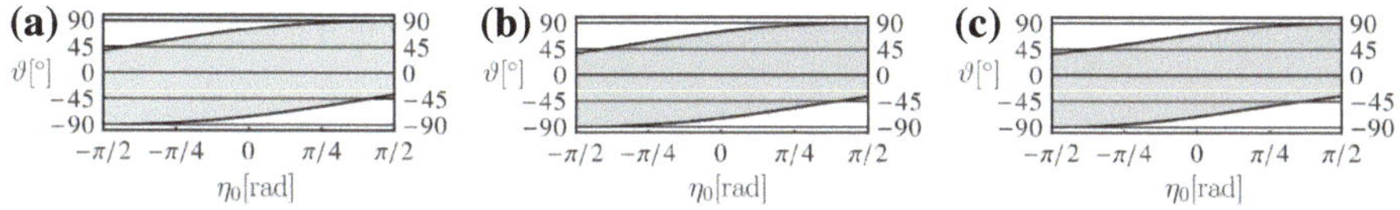

Fig. 3.19 Angular emission ranges according to Eq. (3.60) for the three indicated envelope functions, matched to $N = 2$ and constant peak intensity. **a** $\sin^4$-envelope. **b** Gaussian envelope. **c** Sech-envelope

We fix the width parameter a for both alternative choices by a least square error method to match the $\sin^4$-shaped pulse, what solely ensures that we consider laser pulse of the same duration. The only parameters we consider as given are the laser's amplitude and its central frequency. The resulting shape functions, obtained via that procedure for $\eta_0 = 0$ and $n_C = 2$ in Eq. (2.6) are shown in Fig. 3.18 and we observe little differences between the respective choices for the envelope. The influence of the presented different envelope functions on the emitted radiation's angular distribution is demonstrated in Fig. 3.19. We find the choice of the specific envelope function to be of negligible impact for the applicability of the proposed CEP determination scheme.

Our final remark in this section considers a real limitation for the proposed experimental scheme. We recall that via Eq. (3.60) the CEP determination relies on the fact that the shape functions minimal and maximal value $\psi_{\mathcal{A}}^{\min}(\eta_0)$ and $\psi_{\mathcal{A}}^{\max}(\eta_0)$, respectively, are functions of the CEP. For a long laser pulse, however, this dependency gets lost, since $\psi_{\mathcal{A}}^{\max}(\eta_0) \approx -1$ and $\psi_{\mathcal{A}}^{\max}(\eta_0) \approx 1$. Plotting Eq. (3.58) for different values of n_C in the shape function Eq. (2.6), we can trace the according die out of the angular dependence on η_0. The fact that in Fig. 3.20 the slope of the cutoff angles with respect to η_0 in Fig. 3.20b is negative, whereas in the two other cases it is positive is due to our special choice of the shape function Eq. (2.6). Due to the strong suppression of leading and trailing edge of the pulse, the electron dynamics are dominated by the middle cycles of the laser pulse. In few-cycle pulses the field orientation of these central field cycles experiences a relative sign change, depending on whether we consider an even or odd number of cycles contained in the overall pulse. Thus

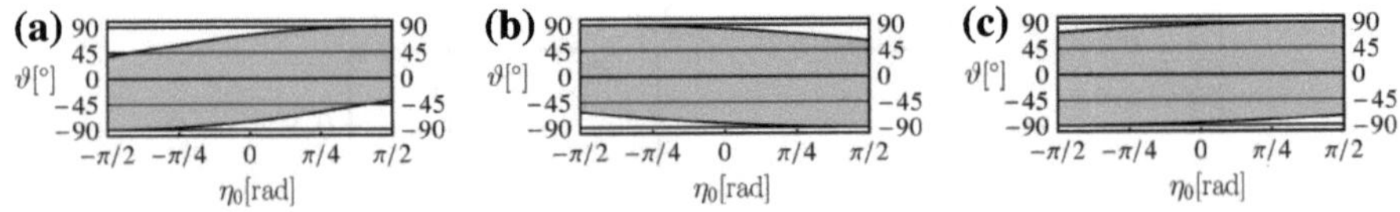

Fig. 3.20 Die out of the η_0-dependence of the angular emission range $n_C \geq 2$ cycles contained in a laser pulse modeled by Eq. (2.6). **a** $n_C = 2$. **b** $n_C = 3$. **c** $n_C = 4$

changing the CEP influences the electron's main emission direction in respectively opposite directions in these both cases. Thus the orientation of the angular emission bands in Fig. 3.20 depends on the specific electron trajectory. Anyway we read off, that for laser pulses comprising more than 5 cycles of the carrying electromagnetic wave, the sketched CEP determination scheme is no longer applicable. In the optical regime this corresponds to laser pulses of little more than 10 fs duration.

3.5 Intensity Dependent Frequency Shift in Pulsed Fields

We consider it in place to elucidate the seemingly elusive concept of mass dressing from the NSCS perspective in the following section of this work. The concept of the so-called dressed mass, originally formulated soon after the formulation of SF-QED in an analysis of the pole structure of the electron propagator in the presence of a strong, oscillating background field [29], and frequently reinvestigated for monochromatic background fields, [30–35], has launched discussion up to the present date [36, 37]. Most of these analyses, considered monochromatic external laser waves, where the mass shift is a clear observation, since the cycle-average of the Dirac current is proportional to the dressed electron momentum, whose square yields an intensity dependent mass shift [see Eqs. (2.115) and (2.119)]. However, rigorously speaking in the analysis of pulsed laser fields quasi-momenta do not appear and thus their square does not correspond to a physical observable. Furthermore since the averaging of the quantum current over many cycles of the laser field is a rather imprecise concept if the field is not strictly periodic, it is desirable to observe effects of the dressed mass in pulsed laser fields. Recently the assertion was raised, that such information can be inferred from the frequency distribution of the radiation of an electron, scattered from a short laser pulse [10]. We want to give a short qualitative discussion, how a dressed mass affects the emission spectra of NSCS in an either monochromatic or pulsed laser field. Comparing Eqs. (1.8) and (2.125) we found that the frequencies emitted from an electron in a monochromatic wave are not equivalent to the usual Doppler shifted frequencies and only in the limit ξ produce to corresponding expression. This intensity dependent frequency shift is due to the increased mass m^*, an electron effectively has inside a monochromatic laser wave. We also note that a similar effect was even apparent in a classical computation [see Eq. (2.44)]. This can be seen by recalling that a photon, emitted in a conventional linear Compton scattering event, which obeys the energy momentum conservation $p_i + k_L = p_f + k_1$, has an energy of

$$\omega_1 = \frac{p_i k_L}{n_1(k_L + p_i)}. \tag{3.66}$$

Changing the electron's momentum by a positive vector along the laser propagation direction $p_i \to p_i + \Delta p\, k_L$ diminishes this emitted photon frequency (please note for this argument $n_1 k_L \geq 0$). This prescription, on the other hand, is precisely the momentum replacement one has to perform in the presence of a monochromatic external laser field, to arrive at the formally equivalent energy momentum conservation $q_i + lk_L = q_f + k_1$, whence we conclude the outgoing photon's frequency to be

$$\omega_1 = \frac{p_i k_L}{n_1((1 + \Delta p_i)k_L + p_i)}. \tag{3.67}$$

The difference between Eqs. (3.66) and (3.67) is called *intensity dependent frequency shift* in nonlinear Compton scattering and can be observed in the scattering of an electron from a monochromatic as well as a pulsed laser field. Its physical explanation is that heavier particles, as an electron with a dressed mass exceeding its rest mass, emit smaller frequencies in Compton scattering. That the frequency shift, obtained in the existing monochromatic analyses is different from that for a few-cycle pulse, is obvious from Fig. 3.2. As, according to Eq. (3.67), the effect of mass dressing can be inferred from changes in the effective electron momentum, in the following we will discuss the average of the quantum current, which is directly connected to the electron's effective momentum in an arbitrary plane wave laser field. It will show, that it is not straightforward to transfer the concept of a dressed momentum to the case of few- cycle laser pulses. This conclusion can be already drawn from the connection of the dressed momentum, with the cycle-averaged quantum current. If the laser field is no longer strictly periodic in all past and future, this average differs from cycle to cycle and a new degree of freedom arises, namely the position of the averaging interval. This freedom of choice already hints at the fact that the dressed mass cannot be universal in pulsed laser fields. On the other hand, if the scattering laser pulse contains many cycles of the carrying wave and thus its spectrum is very narrow, the monochromatic peak positions and thus the corresponding mass dressing reemerges smoothly (see Fig. 3.4). As we consequently wish to make a connection to the monochromatic limit, we are going to compute the averaged quantum current resulting from to the shape function given in Eq. (2.8). We are going to average over the whole invariant phase interval $\eta \in [2\pi(2n_{\text{switch}} + n_{\text{flat}})]$, over which the pulse is non-zero. The Volkov current of a classical electron in the presence of a laser field was given in Eq. (2.105). That expression is averaged over the indicated phase interval by replacing

$$A_L^\mu(\eta) \to 0$$

$$A_L^\mu(\eta)A_{L\mu}(\eta) \to \frac{4\pi^2 n_{\text{switch}}(3n_{\text{flat}} + 2n_{\text{switch}}) - 3\cos(2\eta_0)}{24\pi^2 n_{\text{switch}}(n_{\text{flat}} + 2n_{\text{switch}})} A_L^\mu A_{L\mu}. \tag{3.68}$$

The effective electron momentum resulting from the choice (2.8) for the envelope of the shape function is given by $\langle p^\mu \rangle = \varepsilon V \langle j^\mu \rangle$ and found to be

$$\begin{aligned}\langle p^\mu \rangle &= p_i^\mu + k_L^\mu \frac{e^2 A_L^\mu A_{L\mu}}{4 p_i k_L} \left(\frac{4\pi^2 n_{\text{switch}}(3n_{\text{flat}} + 2n_{\text{switch}}) - 3\cos(2\eta_0)}{12\pi^2 n_{\text{switch}}(n_{\text{flat}} + 2n_{\text{switch}})} \right) \\ &=: p_i^\mu + k_L^\mu \frac{e^2 A_L^\mu A_{L\mu}}{4 p_i k_L} q(n_{\text{flat}}, n_{\text{switch}}, \eta_0). \end{aligned} \tag{3.69}$$

First of all we note that in the limit $n_{\text{flat}} \to \infty$ with fixed n_{switch}, this indeed goes over to the result for a monochromatic, linearly polarized plane wavelength $\langle p^\mu \rangle = q_i^\mu = p_i^\mu + k_L^\mu \frac{e^2 A_L^2}{4(p_i k_L)}$ and this transition is smoothly. For short laser pulses, however, we immediately find new dependencies on the pulse duration $2\pi(n_{\text{flat}} + 2n_{\text{switch}})$, the ratio $n_{\text{flat}}/n_{\text{switch}}$, i. e. the steepness of the pulse's leading and trailing edges, and even on the CEP η_0. The dressed mass, deduced from Eq. (3.69) is given by

$$m^* = \langle p_\mu \rangle \langle p^\mu \rangle = m \sqrt{1 + \frac{\xi^2}{2} q^2(n_{\text{flat}}, n_{\text{switch}}, \eta_0)}, \tag{3.70}$$

where the dependence on the pulse's specific shape and duration is explicit. This kind of ambiguity arises again for the choice of the proper averaging interval. In summary we can say that although all plane wave computations intrinsically contain the monochromatic limit, they do not feature unique patterns of a dressed mass. Averaging phenomena, such as a dressed mass, lose their universality in short laser pulses, but depend delicately on the pulse's specific shape.

References

1. M.E. Peskin, D.V. Schroeder, *An Introduction to Quantum Field Theory* (Westview Press, Boulder, 1995)
2. V.I. Ritus, Quantum effects of the interaction of elementary particles with an intense electromagnetic field. J. Russ. Laser Res. **6**, 497 (1985)
3. C. Harvey, T. Heinzl, A. Ilderton, Signatures of high-intensity Compton scattering. Phys. Rev. A **79**, 063407 (2009)
4. T. Heinzl, D. Seipt, B. Kämpfer, Beam-shape effects in nonlinear Compton and Thomson scattering. Phys. Rev. A **81**, 022125 (2010)
5. A. Ilderton, Trident pair production in strong laser pulses. Phys. Rev. Lett. **106**, 020404 (2011)
6. D. Seipt, B. Kämpfer, Nonlinear Compton scattering of ultrashort intense laser pulses. Phys. Rev. A **83**, 022101 (2011)
7. E.S. Sarachik, G.T. Schappert, Classical theory of the scattering of intense laser radiation by free electrons. Phys. Rev. D **1**, 2738 (1970)

8. Y.I. Salamin, F.H.M. Faisal, Ponderomotive scattering of electrons in intense laser fields. Phys. Rev. A **55**, 3678 (1997)
9. M. Boca, V. Florescu, Nonlinear Compton scattering with a laser pulse. Phys. Rev. A **80**, 053403 (2009)
10. F. Mackenroth, A. Di Piazza, Nonlinear Compton scattering in ultrashort laser pulses. Phys. Rev. A **83**, 032106 (2012)
11. A.H. Compton, A quantum theory of the scattering of x-rays by light elements. Phys. Rev. **21**, 483 (1923)
12. F. Mandl, G. Shaw, *Quantum Field Theory* (Wiley, New York, 1984)
13. L.D. Landau, E.M. Lifschitz, *Lehrbuch der Theoretischen Physik - Band 4: Quantenelektrodynamik* (Akademie Verlag, Berlin, 1991)
14. F. Mackenroth, A. Di Piazza, C.H. Keitel, Determining the carrier-envelope phase of intense few-cycle laser pulses. Phys. Rev. Lett. **105**, 063903 (2010)
15. F.W. Olver, D.W. Lozier, R.F. Boisvert, C.W. Clark, *NIST Handbook of Mathematical Functions* (Cambridge University Press, New York, 2010)
16. F. Krausz, M. Ivanov, Attosecond physics. Rev. Mod. Phys. **81**, 163 (2009)
17. PFS, Petawatt Field Synthesizer http://www.attoworld.de/Home/ourResearch/ToolsAndTechniques/Petawatt-scaleSourceOfIrLight/index.html
18. E. Goulielmakis, M. Uiberacker, R. Kienberger, A. Baltuska, V. Yakovlev, A. Scrinzi, T. Westerwalbesloh, U. Kleineberg, U. Heinzmann, M. Drescher, F. Krausz, Direct measurement of light waves. Science **305**, 1267 (2004)
19. M. Kress, T. Loffler, M.D. Thomson, R. Dorner, H. Gimpel, K. Zrost, T. Ergler, R. Moshammer, U. Morgner, J. Ullrich, H.G. Roskos, Determination of the carrier-envelope phase of few-cycle laser pulses with terahertz-emission spectroscopy. Nat. Phys. **2**, 327 (2006)
20. G.G. Paulus, F. Lindner, H. Walther, A. Baltuska, E. Goulielmakis, M. Lezius, F. Krausz, Measurement of the phase of few-cycle laser pulses. Phys. Rev. Lett **91**, 253004 (2003)
21. T. Wittmann, B. Horvath, W. Helml, M.G. Schaetzel, X. Gu, A. Cavalieri, G.G. Paulus, R. Kienberger, Single-shot carrier-envelope phase measurement of few-cycle laser pulses. Nat. Phys. **5**, 357 (2009)
22. S.P.D. Mangles, C.D. Murphy, Z. Najmudin, A.G.R. Thomas, J.L. Collier, A.E. Dangor, E.J. Divall, P.S. Foster, J.G. Gallacher, C.J. Hooker, D.A. Jaroszynski, A.J. Langley, W.B. Mori, P.A. Norreys, F.S. Tsung, R. Viskup, B.R. Walton, K. Krushelnick, Monoenergetic beams of relativistic electrons from intense laser-plasma interactions. Nature **431**, 535 (2004)
23. C.G.R. Geddes, C. Toth, J. van Tilborg, E. Esarey, C.B. Schroeder, D. Bruhwiler, C. Nieter, J. Cary, W.P. Leemans, High-quality electron beams from a laser wakefield accelerator using plasma-channel guiding. Nature **431**, 538 (2004)
24. J. Faure, Y. Glinec, A. Pukhov, S. Kiselev, S. Gordienko, E. Lefebvre, J.P. Rousseau, F. Burgy, V. Malka, A laser-plasma accelerator producing monoenergetic electron beams. Nature **431**, 541 (2004)
25. W.P. Leemans, B. Nagler, A.J. Gonsalves, C. Toth, K. Nakamura, C.G.R. Geddes, E. Esarey, C.B. Schroeder, S.M. Hooker, GeV electron beams from a centimetre-scale accelerator. Nat. Phys. **2**, 696 (2006)
26. C.E. Clayton, J.E. Ralph, F. Albert, R.A. Fonseca, S.H. Glenzer, C. Joshi, W. Lu, K.A. Marsh, S.F. Martins, W.B. Mori, A. Pak, F.S. Tsung, B.B. Pollock, J.S. Ross, L.O. Silva, D.H. Froula, Self-guided laser wakefield acceleration beyond 1 GeV using ionization-induced injection. Phys. Rev. Lett. **105**, 105003 (2010)
27. G.D. Tsakiris, K. Eidmann, J. Meyer-ter-Vehn, F. Krausz, Route to intense single attosecond pulses. New. J. Phys. **8**, 19 (2006)
28. V. Yanovsky, V. Chvykov, G. Kalinchenko, P. Rousseau, T. Planchon, T. Matsuoka, A. Maksimchuk, J. Nees, G. Cheriaux, G. Mourou, K. Krushelnick, Ultra-high intensity- 300-TW laser at 0.1 Hz repetition rate. Opt. Express. **16**, 2109 (2008)
29. L.S. Brown, T.W.B. Kibble, Interaction of intense laser beams with electrons. Phys. Rev. **133**, A705 (1964)

30. A.I. Nikishov, V.I. Ritus, Quantum processes in the field of a plane electromagnetic wave and in a constant field. 1. Sov. Phys. JETP **19**, 529 (1964)
31. J.H. Eberly, Proposed experiment for observation of nonlinear Compton wavelength shift. Phys. Rev. Lett. **15**, 91 (1965)
32. H.R. Reiss, Proposed experiment to detect the mass shift of an electron in an intense photon field. Phys. Rev. Lett. **17**, 1162 (1966)
33. J.H. Eberly, H.R. Reiss, Electron self-energy in intense plane-wave field. Phys. Rev. **145**, 1035 (1966)
34. J.H. Eberly, A. Sleeper, Trajectory and mass shift of a classical electron in a radiation pulse. Phys. Rev. **176**, 1570 (1968)
35. T.W.B. Kibble, A. Salam, J. Strathdee, Intensity-dependent mass shift and symmetry breaking. Nucl. Phys. B **96**, 255 (1975)
36. C. Harvey, T. Heinzl, A. Ilderton, M. Marklund, The intensity dependent mass shift: existence, universality and detection. Phys. Rev. Lett. **109**, 100402 (2012)
37. T. Heinzl, A. Ilderton, M. Marklund, Finite size effects in stimulated laser pair production. Phys. Lett. B **692**, 250 (2010)

Chapter 4
Nonlinear Double Compton Scattering

A scientist in his laboratory is not only a technician: He is also a child placed before natural phenomena which impress him like a fairy tale.

Marie Curie

4.1 Introduction

A key motivation of this thesis is the fact that SF-QED is just about to enter the nonlinear regime. As we have seen in the last chapter, the quantum probability for NSCS Eq. (3.21) in the classical limit formally reduces to the classical result Eq. (2.17) (see section D.3 for the formal details). On the other hand, it is well known that in the framework of QED the tree level diagrams are merely the lowest order approximation to the exact amplitude. Consequently any process of higher order in the perturbing electromagnetic potential does not feature a formally exact classical analog, as it cannot feature a corresponding classical limit [1]. The inference that in CED one is unable to describe the emission of more than one photon, however, is only partly correct. Of course the question "how many photons" are being emitted, is meaningless in CED, as the radiation field is unquantized (see Fig. 1.1). On the other hand, it was shown by Glauber [2], treating the radiating charge current classically, as in CED, and assuming only the emitted radiation field to be quantized, that the resulting emission spectra relate to the average emitted energy of CED. He could show that, as soon as the computed probability exceeds unity and consequently can no longer be understood as a probability, it can be interpreted as the average number of emitted photons. It is, however, correct to claim that all QED processes of higher than lowest order can not even perturbatively yield exact results of CED. These processes include the emission of any photon number larger than one as well as radiative corrections. Radiative corrections, however, are experimentally only detectable as corrections to

K. F. Mackenroth, *Quantum Radiation in Ultra-Intense Laser Pulses*, Springer Theses, DOI: 10.1007/978-3-319-07740-6_4,

the corresponding fundamental tree-level probabilities. The simplest quantum effect which is experimentally discriminable from the first order process, i.e. does not have a classical analog, thus is the emission of two photons. We refer to this process in the presence of a strong laser field as Nonlinear Double Compton Scattering (NDCS), in analogy to NSCS. From fundamental considerations we would expect that such a two photon emission can occur simultaneously, rendering it a real quantum process, which in the framework of SF-QED is referred to as *coherent* emission. On the other hand, also a classical electron may emit two photons, as just outlined, if its acceleration lasts long enough for two consecutive single photon emissions to become feasible [3]. This process is called *incoherent*. We will comment on the physical meaning of these terms and elaborate on the differences between the two respective processes in Sect. 4.4. In addition to the mentioned analyses of NDCS in the perturbative and monochromatic limit, recently there has also been presented an investigation of NDCS in pulsed external fields, independent of the work presented here [4]. The comparison of the results presented here with those of that work provides a viable check of our calculations.

Since many of the computations of NDCS are, albeit more complex, analogous to those of NSCS, we will frequently refer to the results of Chap. 3 to shorten the discussion. An essential ingredient for computing the scattering matrix element is the electron propagator dressed by the external field, which introduces divergences in case one considers a monochromatic external field (see Sect. 2.2). In the existing analyses of NDCS in such a monochromatic field [5, 6] the authors identified the formally diverging contributions to the scattering matrix element with the incoherent partial processes and considered parameter regimes in which this process was strongly suppressed. They thus eliminated parts of the scattering from their analysis and considered only the non-divergent part. We will see, however, that the contributions to the scattering amplitude, which diverge in the monochromatic limit, not only are non-negligible, but in fact dominate the process of two photon emission. Only in the recent works on NDCS in pulsed external fields [4, 7] a full treatment of the dressed electron propagator in arbitrary plane wave fields was given. In this work we wish to show that the mentioned singularities in the monochromatically dressed electron propagator are only due to the unphysical assumptions of an infinitely long, i.e. single-mode, external plane wave laser field. In Sect. 4.4 they are shown to be absent in pulsed external fields. The NDCS probability integrated over the final phase space of one of the emitted photons gives the differential probability of observing a photon originating from an NDCS event. This quantity is called *inclusive* spectrum and in [4] it was pointed out, that in the regime $\chi \ll 1$, where recoil effects are presumably negligible, it is almost always outscaled by the NSCS emission probability. Hence the authors concluded that in this regime NDCS is difficult to observe since any observed photon is far more likely to stem from NSCS than from NDCS. In Sect. 4.5 we will show that this observation is no longer valid in the full quantum regime $\chi \sim 1$, whence one can think of viable detection schemes of correlated NDCS photons.

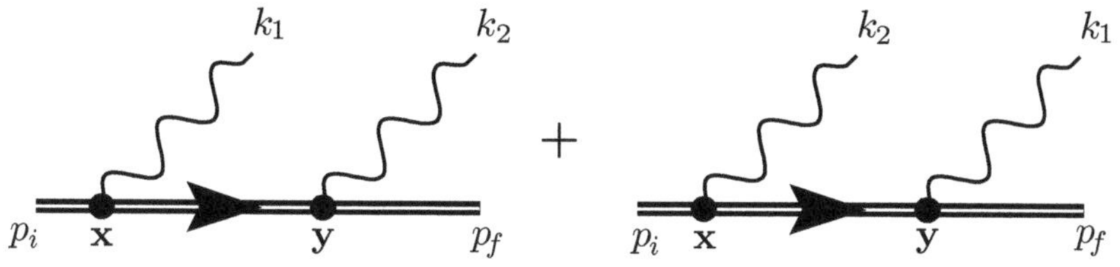

Fig. 4.1 Furry picture feynman diagrams of NDCS

4.2 Matrix Element and Cross Section

We will start our analysis of two photon emission from the second order term of the scattering matrix element, defined in Eq. (2.91), which connects a one electron initial state $|i\rangle = |p_i\rangle$ with a final state containing one electron with changed momentum and two emitted photons $|f\rangle = |p_f; k_1, k_2\rangle$. As is usual in this work we summarize the following technical discussion in a short paragraph at its end, which is then followed by numerical simulations of the analytical results. The according Feynman diagrams are given in Fig. 4.1 and for symmetry reasons it suffices to analyze only one of the partial diagrams and obtain the cross-term by the replacement $1 \leftrightarrow 2$. The analytic expression of the left diagram is given by

$$S_{fi} = -\mathrm{i}\frac{(4\pi)e^2}{\sqrt{16\omega_1\omega_2\varepsilon_i\varepsilon_f V^4}} \int d^4x d^4y\, \overline{\Psi}_{p_f}(y) \not{\epsilon}_2^* \mathrm{e}^{\mathrm{i}k_2 y} G(y,x) \not{\epsilon}_1^* \mathrm{e}^{\mathrm{i}k_1 x} \Psi_{p_i}(x) \tag{4.1}$$

with $\Psi_{p_i}(x)$ ($\overline{\Psi}_{p_f}(x)$) the usual Volkov solution (its conjugate) from Eqs. (2.98) and (2.101) and the dressed electron propagator $G(y, x)$, given in Eq. (2.110). Due to the special space-time dependence of the external plane wave field, all but one coordinate dependency in Eq. (4.1) may be integrated out resulting in

$$\begin{aligned} S_{fi} = &-\mathrm{i}\frac{(2\pi)^4 e^2}{\sqrt{4\omega_1\omega_2\varepsilon_i\varepsilon_f V^4}} \int dx^- dy^- \overline{\Psi}_{p_f}(y^-) \not{\epsilon}_2^* \mathrm{e}^{\mathrm{i}k_2^+ y^-} G(y^-, x^-) \not{\epsilon}_1^* \mathrm{e}^{\mathrm{i}k_1^+ x^-} \Psi_{p_i} \\ &\times \delta^{(3)}(p_i^\perp - k_1^\perp - k_2^\perp - p_f^\perp) \end{aligned} \tag{4.2}$$

with the three perpendicular momentum components $p^\perp$ as defined in Sect. 1.2. The dressed electron propagator with three momentum integrations carried out reads

$$G(y^-, x^-) = \lim_{\epsilon \to 0} \int \frac{dp^+}{2\pi} E_{p_t}(y^-) \frac{\not{p}(p_t^\perp) + m}{p^2(p_t^\perp) - m^2 + i\epsilon} \overline{E}_{p_t}(x^-), \tag{4.3}$$

with the parts of the Ritus matrices depending solely on x^- or y^- from Eq. (2.104). We additionally defined the transitional electron momentum

$$p_t^\mu = \begin{pmatrix} p_t^+ &= \frac{\boldsymbol{p}_t^{\perp 2}+m^2}{2p_t^-} \\ p_t^- &= p_i^- - k_1^- \\ \boldsymbol{p}_t^\perp &= \boldsymbol{p}_i^\perp - \boldsymbol{k}_1^\perp \end{pmatrix}, \tag{4.4}$$

which in addition to the conservation laws for p_t^- and $\boldsymbol{p}_t^\perp$ fulfills the mass-shell condition $p_t^2 = m^2$. Analogous to Eq. (3.16), from the physical requirement $p_t^- > 0$ we infer that quantum mechanically there results a maximally allowed emitted photon frequency

$$\omega_{1,2}^{\mathrm{Max}} = \frac{\varepsilon_i(1+\beta_i)}{1 - \boldsymbol{n}_L \boldsymbol{n}_{1,2}}. \tag{4.5}$$

The notion $p(p_t^\perp)$ in Eq. (4.3) refers to the fact that only p^+ is free, whereas all other three momentum components are fixed to $p_t^\perp$. In Sect. 4.4 it is shown that the dressed propagator can be split up into two contributions of the form

$$\begin{aligned} G(y^-, x^-) &= G^{\mathrm{p.o.}}(y^-, x^-) + G^{\mathrm{s.p.}}(y^-, x^-) \\ G^{\mathrm{p.o.}}(y^-, x^-) &= -\mathrm{i} E_{p_t}(y^-) \frac{\not{p}_t + m}{2p_t^-} \overline{E}_{p_t}(x^-) \Theta(y^- - x^-) \\ G^{\mathrm{s.p.}}(y^-, x^-) &= E_{p_t}(y^-) \frac{\not{k}_L}{2(p_t k_L)} \overline{E}_{p_t}(x^-) \delta(y^- - x^-), \end{aligned} \tag{4.6}$$

with the superscripts standing for *phase ordered* and *same phase*. Inserting this expression now into Eq. (4.2) yields the split up scattering matrix element

$$S_{fi} = -\mathrm{i} \frac{(2\pi)^4 e^2}{\sqrt{4\omega_1\omega_2\varepsilon_i\varepsilon_f V^4}\, 2p_t^-} \left(\widetilde{M}_{fi}^{\mathrm{p.o.}} + \widetilde{M}_{fi}^{\mathrm{s.p.}} \right) \delta^{(3)}(p_i^\perp - k_1^\perp - k_2^\perp - p_f^\perp) \tag{4.7}$$

$$\widetilde{M}_{fi}^{\mathrm{p.o.}} = -\mathrm{i} \sum_{i,j=0}^{2} \widetilde{\Gamma}_{ij} f_{ij}, \quad \widetilde{M}_{fi}^{\mathrm{s.p.}} = \frac{1}{k_{L+}} \sum_{i=0}^{2} \widetilde{\Gamma}_i f_i, \tag{4.8}$$

where the superscripts denote the partial amplitudes resulting from the corresponding term in Eq. (4.6). The partial amplitude $\widetilde{M}_{fi}^{\mathrm{s.p.}}$ in Eq. (4.7) contains dynamic integrals analogous to the definition Eq. (3.5)

$$f_i = \int dx^- \psi_A^i(x^-) \mathrm{e}^{-\mathrm{i} S_{p_i, p_t}^{k_1+k_2}(x^-)} \tag{4.9}$$

$$S^{k_1+k_2}_{p_i,p_f}(x^-) = k_L^+ \int_{-\infty}^{x^-} dc^- \alpha_V\, \psi_{\mathcal{A}}(c^-) + \beta_V\, \psi^2_{\mathcal{A}}(c^-) + \gamma_V \tag{4.10}$$

with the exponential Volkov parameters of the overall process

$$\alpha_V = e\left(\frac{p_i A_L}{k_L p_i} - \frac{p_f A_L}{k_L p_f}\right) \tag{4.11a}$$

$$\beta_V = \frac{e^2 A_L^2}{2}\left(\frac{k_L(k_1+k_2)}{(k_L p_f)(k_L p_i)}\right) \tag{4.11b}$$

$$\gamma_V = -\frac{(k_1+k_2)p_i}{k_L p_f}. \tag{4.11c}$$

The matrix prefactors in Eq. (4.8) are given by

$$\widetilde{\Gamma}_0 = \not{\epsilon}_2^* \not{k}_L \not{\epsilon}_1^*$$

$$\widetilde{\Gamma}_1 = e\left(\frac{\not{A}_L \not{k}_L \not{\epsilon}_1^*(k_L \epsilon_2^*)}{k_L p_f} + \frac{\not{\epsilon}_2^* \not{k}_L \not{A}_L (k_L \epsilon_1^*)}{k_L p_i}\right) \tag{4.12a}$$

$$\widetilde{\Gamma}_2 = -\frac{e^2 A_L^2 (k_L \epsilon_1^*)(k_L \epsilon_2^*)}{(k_L p_i)(k_L p_f)} \not{k}_L \tag{4.12b}$$

Again as in the case of NSCS (see Chap. 3) the function f_0 is divergent. Its regularization is formally equivalent to Eq. (3.13). The partial amplitude $\widetilde{M}^{\text{p.o.}}_{fi}$ in Eq. (4.7) differs more strongly from the NSCS analysis. It is governed by two dimensional dynamic integrals of the form

$$f_{ij} = \int dx^- dy^- \Theta(y^- - x^-) \psi^j_{\mathcal{A}}(x^-) \psi^i_{\mathcal{A}}(y^-) \mathrm{e}^{-\mathrm{i}\,(S^{k_1}_{p_i,p_t}(x^-) + S^{k_2}_{p_t,p_f}(y^-))} \tag{4.13}$$

containing the actions

$$S^{k_1}_{p_i,p_t}(x^-) = g^{k_1}_{p_i,p_t}(x^-) + k_L^+ \gamma_V^x \tag{4.14a}$$

$$S^{k_2}_{p_t,p_f}(y^-) = g^{k_2}_{p_t,p_f}(y^-) + k_L^+ \gamma_V^y$$

$$g^{k_1}_{p_i,p_t}(x^-) = k_L^+ \int_{-\infty}^{x^-} dc^-\ \alpha_V^x\, \psi_{\mathcal{A}}(c^-) + \beta_V^x\, \psi^2_{\mathcal{A}}(c^-)$$

$$g^{k_2}_{p_t,p_f}(y^-) = k_L^+ \int_{-\infty}^{y^-} dc^-\ \alpha_V^y\, \psi_{\mathcal{A}}(c^-) + \beta_V^y\, \psi^2_{\mathcal{A}}(c^-) \tag{4.14b}$$

which also contain Volkov-type exponential parameters. In this case, however, there are two sets of such parameters, one for each interaction vertex, given by

$$\alpha_V^x = e\left(\frac{p_i A_L}{p_i k_L} - \frac{p_t A_L}{p_t k_L}\right), \qquad \alpha_V^y = e\left(\frac{p_t A_L}{p_t k_L} - \frac{p_f A_L}{p_f k_L}\right) \tag{4.15a}$$

$$\beta_V^x = \frac{e^2 A_L^2}{2}\left(\frac{k_1 k_L}{(p_i k_L)(p_t k_L)}\right), \quad \beta_V^y = \frac{e^2 A_L^2}{2}\left(\frac{k_2 k_L}{(p_t k_L)(p_f k_L)}\right) \tag{4.15b}$$

$$\gamma_V^x = -\frac{k_1 p_i}{p_t k_L}, \qquad \gamma_V^y = -\frac{k_2 p_t}{p_f k_L}. \tag{4.15c}$$

By definition these parameters fulfill the relations

$$\alpha_V^x + \alpha_V^y = \alpha_V \tag{4.16a}$$

$$\beta_V^x + \beta_V^y = \beta_V \tag{4.16b}$$

$$\gamma_V^x + \gamma_V^y = \gamma_V, \tag{4.16c}$$

$$S_{p_i,p_t}^{k_1}(x^-) + S_{p_t,p_f}^{k_2}(x^-) = S_{p_i,p_f}^{k_1+k_2}(x^-). \tag{4.16d}$$

The prefactors of these bivariate parameter functions are given by

$$\widetilde{\Gamma}_{ij} = \widetilde{\Gamma}_i^2\left[\not{p}_t + m\right]\widetilde{\Gamma}_j^1$$

$$\widetilde{\Gamma}_0^l = \not{\epsilon}_l^*$$

$$\widetilde{\Gamma}_1^1 = e\left(\frac{\not{A}_L \not{k}_L \not{\epsilon}_l^*}{2(k_L p_i)} + \frac{\not{\epsilon}_l^* \not{k}_L \not{A}_L}{2(k_L p_t)}\right), \quad \widetilde{\Gamma}_1^2 = e\left(\frac{\not{A}_L \not{k}_L \not{\epsilon}_l^*}{2(k_L p_t)} + \frac{\not{\epsilon}_l^* \not{k}_L \not{A}_L}{2(k_L p_f)}\right)$$

$$\widetilde{\Gamma}_2^1 = -\frac{e^2 A_L^2 (k_L \epsilon_1^*)}{2(k_L p_i)(k_L p_t)}\not{k}_L, \qquad \widetilde{\Gamma}_2^2 = -\frac{e^2 A_L^2 (k_L \epsilon_2^*)}{2(k_L p_t)(k_L p_f)}\not{k}_L \tag{4.17}$$

In the parameter functions f_{ij} one faces the same difficulty as in the case of the f_i. Those integrals where one of the integration dimensions is not temporally confined by the shape function in the preexponent (i.e. $i, j = 0$), are divergent. To deal with this difficulty a similar trick as in the previous case needs to be performed. To this end we define two inner integral functions

$$\begin{aligned} f_{ij} &= \int_{-\infty}^{\infty} dy^- dx^- \Theta(y^- - x^-)\psi_{\mathcal{A}}^j(x^-)\psi_{\mathcal{A}}^i(y^-)\mathrm{e}^{-\mathrm{i}\,(S_{p_i,p_t}^{k_1}(x^-)+S_{p_t,p_f}^{k_2}(y^-))} \\ &=: \int_{-\infty}^{\infty} dy^-\, \mathbb{G}^j(y^-)\psi_{\mathcal{A}}^i(y^-)\mathrm{e}^{-\mathrm{i}\,S_{p_t,p_f}^{k_2}(y^-)} \\ &=: \int_{-\infty}^{\infty} dx^-\, \mathbb{I}^i(x^-)\psi_{\mathcal{A}}^j(x^-)\mathrm{e}^{-\mathrm{i}\,S_{p_i,p_t}^{k_1}(x^-)}. \end{aligned} \tag{4.18}$$

For these inner integral functions it holds

$$\begin{aligned}\partial_{y^-}\mathbb{G}^j(y^-) &= \psi^j_{\mathcal{A}}(y^-)\mathrm{e}^{-\mathrm{i}\,S^{k_1}_{p_i,p_t}(y^-)}\\ \partial_{x^-}\mathbb{I}^i(x^-) &= -\psi^i_{\mathcal{A}}(x^-)\mathrm{e}^{-\mathrm{i}\,S^{k_2}_{p_t,p_f}(x^-)}.\end{aligned} \tag{4.19}$$

We further note that by integration by parts the following relations hold

$$\begin{aligned}\int_{-\infty}^{\infty} dy^-\left(\partial_{y^-}\mathbb{G}^j(y^-)\right)\mathrm{e}^{-\mathrm{i}\,S^{k_2}_{p_t,p_f}(y^-)} &= -\int_{-\infty}^{\infty} dy^-\,\mathbb{G}^j(y^-)\partial_{y^-}\left(\mathrm{e}^{-\mathrm{i}\,S^{k_2}_{p_t,p_f}(y^-)}\right)\\ \int_{-\infty}^{\infty} dx^-\left(\partial_{x^-}\mathbb{I}^i(x^-)\right)\mathrm{e}^{-\mathrm{i}\,S^{k_1}_{p_i,p_t}(x^-)} &= -\int_{-\infty}^{\infty} dx^-\,\mathbb{I}^i(x^-)\partial_{x^-}\left(\mathrm{e}^{-\mathrm{i}\,S^{k_1}_{p_i,p_t}(x^-)}\right).\end{aligned} \tag{4.20}$$

Then by use of Eq. (4.16d) it follows

$$\gamma^y_V\, f_{0j} = -\mathrm{i}f_j - \left(\alpha^y_V\, f_{1j} + \beta^y_V\, f_{2j}\right) \tag{4.21a}$$

$$\gamma^x_V\, f_{i0} = \mathrm{i}f_i - \left(+\alpha^x_V\, f_{i1} + \beta^x_V\, f_{i2}\right). \tag{4.21b}$$

Analogous to Eq. (3.15) these expressions can be equally understood as manifestations of gauge invariance [4, 8]. Applying Eqs. (4.21a) and (4.21b) at once one finds

$$\begin{aligned}f_{00} = &\frac{\mathrm{i}}{\gamma^x_V\gamma^y_V}\left(\frac{\gamma^x_V\alpha^x_V - \gamma^y_V\alpha^y_V}{\gamma_V}f_1 + \frac{\gamma^x_V\beta^x_V - \gamma^y_V\beta^y_V}{\gamma_V}f_2\right)\\ &+\frac{\mathrm{i}}{\gamma^x_V\gamma^y_V}\left(\alpha^x_V\alpha^y_V\, f_{11} + \alpha^x_V\beta^y_V\, f_{21} + \beta^x_V\alpha^y_V\, f_{12} + \beta_{V1}\beta^y_V\, f_{22}\right).\end{aligned} \tag{4.22}$$

Now since the dynamics of the process is captured in the parameters functions, from the above formulas we see that portions of the partial matrix element $\widetilde{M}^{\text{p.o.}}_{fi}$ are governed by univariate parameter functions and thus physically need to be attributed to $\widetilde{M}^{\text{s.p.}}_{fi}$. The full scattering matrix element with all divergences removed and split up into physical subprocesses thus reads

$$\begin{aligned}S_{fi} &= \mathcal{N}_{\text{NDCS}}\,\bar{u}_{p_f}\left(M^{\text{p.o.}}_{fi} + M^{\text{s.p.}}_{fi}\right)u_{p_i}\delta^{(3)}(p_i^{\perp} - k_1^{\perp} - k_2^{\perp} - p_f^{\perp})\\ \mathcal{N}_{\text{NDCS}} &= -\mathrm{i}\frac{(2\pi)^4 e^2}{\sqrt{4\omega_1\omega_2\varepsilon_i\varepsilon_f V^4}}\end{aligned}$$

$$M_{fi}^{\text{p.o.}} = -\frac{\text{i}}{2p_t^-}\sum_{i,j=1}^{2}\Gamma_{ij}\, f_{ij}, \qquad M_{fi}^{\text{s.p.}} = \frac{1}{2(k_L p_t)}\sum_{i=1}^{2}\Gamma_i\, f_i \tag{4.23}$$

The intricacies of connecting these partial amplitudes to physically discriminable partial processes of NDCS, as discussed at the beginning of this chapter, are discussed in Sect. 4.4. The coefficients of the physical same phase matrix element $M_{fi}^{\text{s.p.}}$ are expressed in terms of Eqs. (4.12) and (4.17) according to

$$\Gamma_1 = \widetilde{\Gamma}_1 - \frac{\alpha_V}{\gamma_V}\widetilde{\Gamma}_0 + \frac{\alpha_V^x\gamma_V^x - \alpha_V^y\gamma_V^y}{2\gamma_V\gamma_V^x\gamma_V^y}\Gamma_{00} - \frac{1}{\gamma_V^x}\Gamma_{10} + \frac{1}{\gamma_V^y}\Gamma_{01} \tag{4.24a}$$

$$\Gamma_2 = \widetilde{\Gamma}_2 - \frac{\beta_V}{\gamma_V}\widetilde{\Gamma}_0 + \frac{\beta_V^x\gamma_V^x - \beta_V^y\gamma_V^y}{2\gamma_V\gamma_V^x\gamma_V^y}\Gamma_{00} - \frac{1}{\gamma_V^x}\Gamma_{20} + \frac{1}{\gamma_V^y}\Gamma_{02}. \tag{4.24b}$$

The definition of the matrix coefficients of the phase ordered partial amplitude Γ_{ij}, given in Eq. (4.17), changes accordingly. To obtain a physical quantity such as the differential emission probability we have to take the modulus square of the scattering matrix element, according to the procedure outlined in Appendix A. The resulting differential probability of an electron upon interaction with a pulsed plane wave field to emit two photons is then given by

$$dW = \frac{e^4\,\omega_1\omega_2}{4(2\pi)^4 p_i^- p_f^-}\left|\bar{u}_{p_f}\left(M_{fi}^{\text{s.p.}} + M_{fi}^{\text{p.o.}}\right)u_{p_i} + (1 \leftrightarrow 2)\right|^2 d\omega_1 d\Omega_1 d\omega_2 d\Omega_2, \tag{4.25}$$

where $(1 \leftrightarrow 2)$ denotes the cross channel term, corresponding to the Feynman diagram shown right in Fig. 4.1, which is obtained from Eq. (4.23) by exchanging the two numeric indices 1 and 2. This probability is differential in the six quantities $\omega_{1,2}$ and $(\vartheta_{1,2}, \varphi_{1,2})$. Thus, whenever we are going to present numerical simulations of Eq. (4.25) we are going to fix four of these values, e.g. the observation directions of the two photons $(\vartheta_{1,2}, \varphi_{1,2})$, and show a two-fold differential probability in the remaining quantities. It is interesting to note that the partial amplitude $S_{fi}^{\text{p.o.}}$ of Eq. (4.23) can be interpreted written in terms of two separate NSCS scattering amplitudes. For the following discussion we will again focus only on the channel of NDCS depicted left in Fig. 4.1. An analogous result holds for the cross channel. To demonstrate the connection to two separate NSCS amplitudes, we denote the NSCS amplitude for an electron of initial momentum p_i^μ to emit a photon of wave vector k_1^μ, thereby changing its momentum to p_t^μ, as $S_{fi}^{\text{NSCS}}\,(p_i \to p_t, k_1)$ and the matrix coefficients of the reduced matrix element Eq. (3.4) are analogously labeled $\Gamma_j^{\text{NSCS}}\,(p_i \to p_t, k_1)$. From Eq. (3.3) we then derive the product of two NSCS amplitudes for the successive emission of two photons with wave vectors $k_{1,2}^\mu$ via the transitional electron momentum state p_t^μ

$$S_{fi}^{\text{NSCS}}\left(p_t \to p_f, k_2\right) \times S_{fi}^{\text{NSCS}}\left(p_i \to p_t, k_1\right)$$
$$= -\frac{e^2(2\pi)^7}{\sqrt{16\omega_1\omega_2\varepsilon_i\varepsilon_f V^6\varepsilon_t}}\delta^{(3)}\left(p_i^{\perp} - k_1^{\perp} - p_t^{\perp}\right)\delta^{(3)}\left(p_t^{\perp} - k_2^{\perp} - p_f^{\perp}\right)$$
$$\times \int dy^- \sum_{i=1}^{2} \bar{u}_{p_f}\Gamma_i^{\text{NSCS}}\left(p_t \to p_f, k_2\right)\psi_{\mathcal{A}}^i(y^-)u_{p_t,\sigma_t}\mathrm{e}^{\mathrm{i}S_{p_t,p_f}^{k_2}(y^-)}$$
$$\times \int dx^- \sum_{j=1}^{2} \bar{u}_{p_t,\sigma_t}\Gamma_j^{\text{NSCS}}\left(p_i \to p_t, k_1\right)\psi_{\mathcal{A}}^j(x^-)u_{p_i}\mathrm{e}^{\mathrm{i}S_{p_i,p_t}^{k_1}(x^-)}, \tag{4.26}$$

where σ_t indicates the spin states of the electron in the transitional momentum state p_t^{μ}. Using the completeness relation Eq. (B.18) for the spinors $u_{p_t,\sigma}$, the latter expression is easily connected with the phase ordered partial scattering matrix element from Eq. (4.23)

$$S_{fi}^{\text{p.o.}} = \sum_{\sigma_t}\int dx^- dy^- \frac{d^3 p_t^{\perp} V}{(2\pi)^3}\Theta(y^- - x^-)$$
$$\times \left(\frac{\partial}{\partial y^-}S_{fi}^{\text{NSCS}}\left(p_t \to p_f, k_2\right)\right)\left(\frac{\partial}{\partial x^-}S_{fi}^{\text{NSCS}}\left(p_i \to p_t, k_1\right)\right). \tag{4.27}$$

In this expression $(d^3 p_t^{\perp} V)/(2\pi)^3$ is the phase space volume of the intermediate electron. From Eq. (4.27) we thus read off that the partial scattering amplitude proportional to $M_{fi}^{\text{p.o.}}$ corresponds to an electron with initial momentum p_i^{μ} undergoing the process of emitting a photon with wave vector k_1^{μ} at vertex x, changing its four momentum to p_t^{μ}, and subsequently emitting a second photon with wave vector k_2^{μ} at vertex y, changing it four momentum to p_f^{μ}. It is interesting to note that the step function, which naturally occurred during our computation, in view of this interpretation simply reflects causality. Since we computed the partial amplitude of the process depicted left in Fig. 4.1 the photon with four momentum k_1 needs to be emitted before k_2. This is reflected in the step function, which gives a zero contribution unless the phase of the second vertex is larger than that of the first one. The phase ordered partial matrix element $S_{fi}^{\text{p.o.}}$ thus corresponds to a time ordered succession of two single Compton events, summed over all spin and momentum states of the intermediate electron. The same phase contribution does not feature a comparable decomposition into two subprocesses. But it is connected to the perturbative double Compton effect [9, 10]. However, usual electron-photon scattering is a single-photon process. The laser field thus needs to be weak, to recover this limit, so that the electron will interact with only one photon from it. Thus the partial amplitude $S_{fi}^{\text{s.p.}}$ is connected to the perturbative limit of the full scattering matrix element. The rather technical formal details of this as well as the monochromatic limit of the general NDCS emission probability Eq. (4.25) are given in Appendix E.

4.2.1 Summarizing the Above Discussion

We have analytically investigated a particular channel of NDCS, corresponding to the Feynman diagram depicted left in Fig. 4.1. The cross channel, depicted right in Fig. 4.1, is obtained from our results by the replacement $1 \leftrightarrow 2$. We demonstrated how the scattering matrix element naturally decomposes into two partial amplitudes, by virtue of a split up of the dressed electron propagator, which is discussed in Sect. 4.4. We labeled these two partial amplitudes as the *same phase* and *phase ordered* contribution, as they are governed by dynamic integrals which depend on only one or on two causally ordered phase variables, respectively. Causal ordering in this respect means that in the phase ordered contribution to the NDCS channel depicted left in Fig. 4.1 the photon with wave vector k_1^μ has to be emitted before that with wave vector k_2^μ. Finally we have demonstrated that the phase ordered contribution to NDCS is equivalent to a phase ordered succession of two NSCS scattering amplitudes, which further supports the assertion that the photon emission is causally structured.

For the further evaluation of Eq. (4.25), just as in the case of NSCS, we employ a numerical integration scheme of the dynamic integrals (see Chap. 5). The algebraic spin summation and average are also performed numerically. Throughout this chapter we are going to consider the interaction of an electron and a laser pulse in a reference frame as depicted in Fig. 2.1. The first step is now to seek for confirmation of the correctness of the analytic as well as numerical results. As outlined in the opening of this chapter, the comparison of the results obtained by a numerical simulation of Eq. (4.25) to those of an independent analysis of NDCS can be employed as such a consistency check. In Fig. 4.2a[1] we show the probability of an electron with initial energy $\varepsilon_i/m = 10^4$ (corresponding to $\varepsilon_i \approx 5$ GeV), colliding head on with a plane wave laser pulse of central frequency $\omega_L = 1.55$ eV and intensity $I_L = 2 \times 10^{18} (\xi = 1)$, to emit two photons into the directions $\vartheta_{1,2} = m/\varepsilon_i$, $\varphi_1 = \pi/2$ and $\varphi_2 = 3\pi/2$ as a function of ω_1, ω_2, published in [4]. In Fig. 4.2b we show the result of a numerical simulation of Eq. (4.25) in comparison. Smaller deviations between the two respective spectra result from a difference in the employed shape functions. In [4] the authors employ a pulse form $\psi_\mathcal{A}(\eta) = \cosh(\pi\eta/2\tau)\cos(\eta)$ for $-\tau \leq \eta \leq \tau$ and zero elsewhere, together with a dimensionless pulse length of $\tau = 20$. To recover the resulting shape function from our pulse model Eq. (2.6) we employ $n_C = 20/\pi \approx 6$, whence we derive the same pulse duration. From a comparison of the resulting shape functions, shown in Fig. 4.2c, we read off a different rise and fall of the shape function's edges, which is due to the different power law scaling of the pulse envelope in Eq. (2.6) as compared to that one employed in [4]. In view of these differences the good qualitative agreement between Fig. 4.2a, b is striking. In Sect. E.3 it is demonstrated that in accordance with [11] the scattering amplitude exhibits resonances at the frequencies.

---

[1] Reprinted figure with permission from [4]. Copyright (2012) by the American Physical Society.

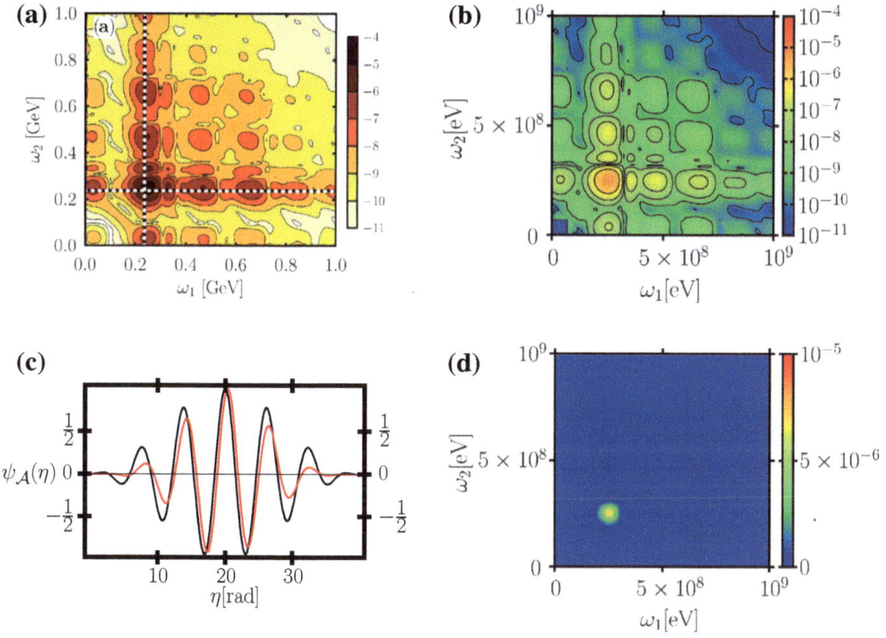

Fig. 4.2 Comparison of a numerical simulation of Eq. (4.25) to the results published in [4]. The emission probabilities are given in [(eV sr)$^{-2}$]. **a** Two photon emission probability $dW/(\prod_{i=1,2} d\omega_i d\Omega_i)$. The *dashed lines* give the position of the first Oleinik resonance in the emission spectra. They are well reproduced in Fig. 4.2b. **b** Two photon emission probability $dW/(\prod_{i=1,2} d\omega_i d\Omega_i)$ computed from Eq. (4.25). **c** Shape functions employed in [4] (*black*, shifted in η-direction by τ) and derived from Eq. (2.6) for a cycle number $n_C = 20/\pi$ (*red*). **d** Emission probability of Fig. 4.2b on a linear scale

$$\omega_{1,2}^{\mathrm{m.c.},l} = \frac{l(p_i k_L)}{(p + (l + \frac{m^2\xi^2}{4(p_i k_L)})k_L)n_{1,2}}. \tag{4.28}$$

The first resonances in ω_1 and ω_2 are marked by dotted lines in Fig. 4.2a. Due to the fact that the first maximum well reproduces this monochromatic result, we conclude that for a the pulse duration of $n_C \approx 6$ the assumption of a singular frequency spectrum yields adequate results. We point out that the higher order resonance maxima appearing in both $\omega_{1,2}$-directions are caused by the electron interacting with more than one photon from the laser field, but nevertheless they are not significant for the overall emission probability shown in Fig. 4.2. In fact, showing the emission probability on a linear scale (see Fig. 4.2d), we find only the first harmonic to be of detectable significance. The observed good agreement with the monochromatic case, at least concerning the position of the resonance peaks, is not persistent for arbitrary pulse shapes. In the case of NSCS in section D.3 we found that for the choice $n_C = 2$ in Eq. (2.6), already for $\xi \lessapprox 1$ the NSCS result significantly differs from a monochromatic analysis. To study this regime, where the monochromatic approximation is no longer valid, we will next consider a scenario where the laser pulse is modeled by Eq. (2.6) with $n_C = 2$ cycles contained. Furthermore we aim at comparing the respective magnitudes of the two separate contributions $S_{fi}^{\text{s.p.}}$ and $S_{fi}^{\text{p.o.}}$ to the overall scattering process. To this end we simply square each of these partial amplitudes separately in Eq. (4.25) and label the resulting emission probabilities $dW^{\text{s.p.}}$ and $dW^{\text{p.o.}}$. We stress, that in order to resemble an experimental measurement, we are not going to show the emission probabilities, but the corresponding energy emission spectra, according to the discussion of appendix A. We study the emission from an in our terms moderately relativistic electron with an initial energy of $\varepsilon_i = 100$ MeV, scattered from a laser pulse with an intensity of $I_L = 10^{20}\text{W/cm}^2$ (corresponding to $\xi \approx 7$). We choose to observe the radiation at $\vartheta_{1,2} = \pi - \vartheta_0/2$, $\varphi_{1,2} = 0$. We use $\vartheta_0 = m\xi/\varepsilon_i$ in accordance to the discussion of Sect. 3.4. In Fig. 4.3 we show the spectra resulting from the according simulation of Eq. (4.25). In Fig. 4.3c, d we show the same spectra as in Fig. 4.3a, b, albeit with a logarithmic scaling of the emitted photon frequencies $\omega_{1,2}$. Presenting the emission spectra in this way allows for a simultaneous investigation of the different relative frequency regimes $\omega_1 \sim \omega_2$ and $\omega_{1,2} \gg \omega_{2,1}$ where $\omega_{2,1}$ has to be understood as the respectively other frequency than that on the left side of this inequality. Since we will find important features of the scattering process in the latter frequency regimes, we are going to show some logarithmic spectra and it is instructive to have an exemplary comparison between a logarithmic and the corresponding linear spectrum at hand, for the sake of physical intuition. We draw attention to three observable main features in Fig. 4.3: First we have indicated the frequency $\omega_1^{\text{m.c.},100}$, where the 100th Oleinik resonance as obtained from Eq. (4.28) is located, by a white vertical line. We note that this significantly large monochromatic resonance order is located between the second and third resonance peaks, found from the analysis of a pulsed field. This shift of the resonance peaks again is explainable as a reduced red-shift due to a decreased mass dressing in the few-cycle laser pulse (see Sect. 3.5). The monochromatic analysis is thus also in the case of NDCS found to be of no predictive power in the few-cycle regime, as also observed in Chap. 3. Next, comparing Figs. 4.3a, c to 4.3b, d we find a suppression of $dE^{\text{s.p.}}$ with respect to the phase ordered contribution by five orders of magnitude. We can consequently focus the following analysis of NDCS on the phase ordered partial process, which will lead to simple physical pictures. Thirdly, in Fig. 4.3b, we note the regular distribution of the resonance peaks in the two emitted photons' frequencies. Recalling Eq. (4.27), this structure is strongly reminiscent of the peak structure we found for NSCS spectra (compare Fig. 3.2b). Denoting the probability of an electron with initial momentum p_i^μ to emit two photons with wave vectors $k_{1,2}^\mu$ via NDCS by $dW_{p_i}^{\text{NDCS}}(k_1, k_2)$ and the probability of an electron to emit a single photon with wave vector $k_{1,2}^\mu$ in an NSCS event as $dW_{p_i}^{\text{NSCS}}(k_{1,2})$, the spectra found in Fig. 4.3a, d then have the formal structure

$$dW_{p_i}^{\text{NDCS}}(k_1, k_2) = dW_{p_i}^{\text{NSCS}}(k_1) \times dW_{p_i}^{\text{NSCS}}(k_2). \tag{4.29}$$

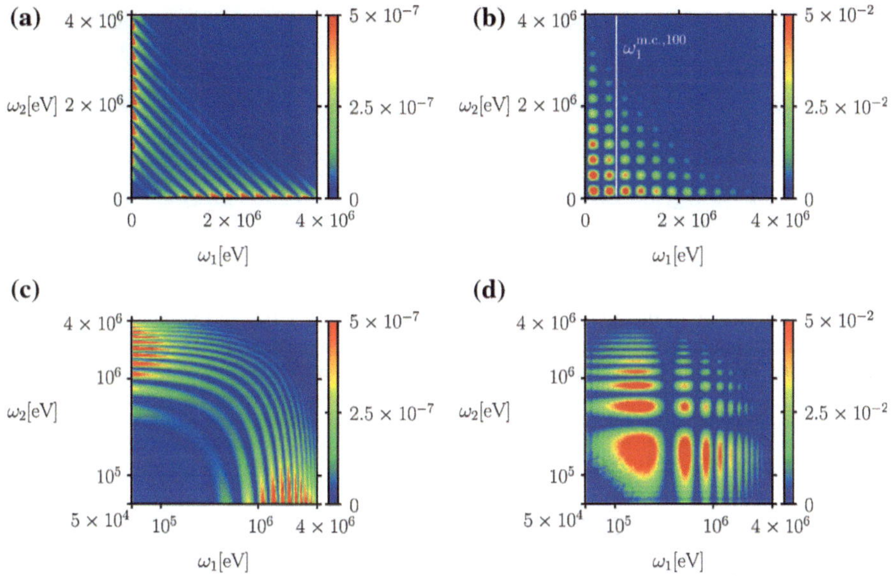

Fig. 4.3 Comparison of the energy emission spectra $dE/(\prod_{i=1,2} d\omega_i d\Omega_i)$ resulting from the two partial contributions to the emission of two photons, observed at $\vartheta_{1,2} = \pi - \vartheta_0/2$, $\varphi_{1,2} = 0$, emitted from an electron with initial energy $\varepsilon_i = 100\,\text{MeV}$, scattered from a laser pulse with intensity $I_L = 10^{20}\,\text{W cm}^{-2}$. The spectra are given in [$\text{eV}^{-1}\ \text{sr}^{-2}$]. **a** Same phase contribution. **b** Phase ordered contribution. The white horizontal line indicates the 100th Oleinik resonance $\omega_1^{\text{m.c.},100}$. **c** Same phase contribution with logarithmic frequency axes. **d** Phase ordered contribution with logarithmic frequency axes

The neglect of the same phase contribution in this argument is in agreement with the strong suppression found in Fig. 4.3. The two NSCS emission events, constituting the phase ordered contribution, and thus approximately the whole probability of an NDCS event, will be labeled *independent* of each other, if the NDCS emission probability can be approximated according to Eq. (4.29) with the initial momenta of both NSCS events the same as the initial momentum of the NDCS event. Please note that Eq. (4.29) is fundamentally different from Eq. (4.27) in as much as the initial momenta of the two NSCS probabilities are the same. If this is not the case, then Eq. (4.27) still holds, in Eq. (4.29), on the other hand, one would have to insert a changed initial momentum for the second NSCS probability. This initial momentum of the second emission event, on the other hand, will depend on the photon energy emitted in the first NSCS event, whence the two events can no longer be called independent. We confirm that the two NSCS events constituting the NDCS spectrum of Fig. 4.3a indeed are independent by graphically comparing this spectrum to a spectrum obtained according to the right hand side of Eq. (4.29), shown in Fig. 4.4b. In fact, we find an excellent agreement. The different color scales in Figs. 4.3a and 4.4a are due to the fact that in the former we show the energy emitted, whereas in the latter only the emission probabilities are given. These quantities differ by a

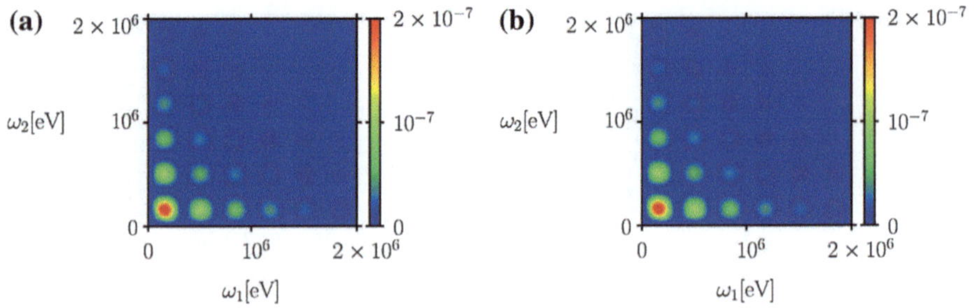

Fig. 4.4 Comparison of the NDCS emission probability as a function of the emitted photon frequencies in comparison to the product of two NSCS emission probabilities, multiplied according to Eq. (4.29). The emission probabilities are given in [(eV sr)$^{-2}$]. **a** Total emission probability $dW_{p_i}^{\text{NDCS}}(k_1, k_2)$. **b** Product of the two NSCS probabilities obtained from Eq. (3.21) for the parameters of Fig. 4.3a

factor $(\omega_1 + \omega_2)$. We resort to giving these two different quantities, since the natural quantity to be measured in an experiment is the emitted energy, which on the other hand does not lend itself to a product expression analogous to Eq. (4.29).

Having established these results for the general structure of NDCS emission probabilities, we note that as in the case of NSCS the dynamic integrals can be analytically approximated in various asymptotic scenarios. As we did in Chap. 3 we are going to omit the technically involved details of these limits to the appendix. In particular we study the perturbative limit of NDCS in section E.2 as well as its monochromatic limit in section E.3. The approximation of the dynamic integrals in the relativistic regime $m\xi, \varepsilon_i \gg m$, however, will again be explicitly studied in the following section.

4.3 Stationary Phase Approximation

Just as in the case of NSCS it is highly beneficial to obtain a simple approximation scheme for the NDCS scattering amplitude in the relativistic regime $m\xi, \varepsilon_i \gg m$. To find such a scheme we again have to analyze the dynamic integrals defined in Eqs. (4.9) and (4.13). In this respect we note that the exponentials of the univariate dynamic integrals f_i are given by differences of classical actions which are formally equivalent to Eq. (3.12). For these integrals then Eqs. (3.36), (3.43), (3.47) and (3.48) can be unchangeably transferred. The accordingly only remaining question is how the bivariate dynamic integrals Eq. (4.13) can be subjected to a stationary phase analysis. Along the lines of the argumentation of Sect. 3.3 we conclude that in the relativistic regime the exponential phases of the bivariate dynamic integrals are large, their integrands thus rapidly oscillating and that they consequently are accessible to a stationary phase approximation. From the general theory of the multidimensional saddle point approximation it follows, that for a bivariate integrand, as realized in Eq. (4.13), a stationary point is distinguished by the condition

$$\nabla_{x^-,y^-}\left(S^{k_1}_{p_i,p_t}(x^-)+S^{k_2}_{p_t,p_f}(y^-)\right)=0, \tag{4.30}$$

where $\nabla_{x^-,y^-}=(\partial_{x^-},\partial_{y^-})$ is the vector of the partial derivatives. From the decoupling of the phases in the two integration dimensions x^-, y^-, apparent in Eq. (4.30) and from Eq. (4.14) we read off, that the stationary points of the bivariate dynamic integrals are pairs of stationary points $(\mathring{x}^{-,l},\mathring{y}^{-,k})$, which are distinguished by conditions analogous to Eq. (3.23)

$$\alpha^x_V\psi_\mathcal{A}(\mathring{x}^{-,l})+\beta^x_V\psi^2_\mathcal{A}(\mathring{x}^{-,l})+k^+_L\gamma^x_V=0 \tag{4.31a}$$
$$\alpha^y_V\psi_\mathcal{A}(\mathring{y}^{-,k})+\beta^y_V\psi^2_\mathcal{A}(\mathring{y}^{-,k})+k^+_L\gamma^y_V=0, \tag{4.31b}$$

We note that according to the discussion from Sect. 3.3 these equations only feature a real solution if the following conditions are satisfied

$$\frac{(m\xi)}{\varepsilon_i}\psi^{\min}_\mathcal{A}\le(\beta_i\epsilon_L)+(\beta_ik_L)\frac{k_1\epsilon_L}{k_1k_L}\le\frac{(m\xi)}{\varepsilon_i}\psi^{\max}_\mathcal{A} \tag{4.32a}$$
$$\frac{(m\xi)}{\varepsilon_t}\psi^{\min}_\mathcal{A}\le(\beta_t\epsilon_L)+(\beta_tk_L)\frac{k_2\epsilon_L}{k_2k_L}\le\frac{(m\xi)}{\varepsilon_t}\psi^{\max}_\mathcal{A}, \tag{4.32b}$$

where again the four velocities $\beta^\mu=p^\mu/\varepsilon$ are inserted. Since we only integrate over the half space $y^->x^-$ contributions from stationary point pairs with $\mathring{y}^{-,k}<\mathring{x}^{-,l}$ do not contribute to our analysis. Assume now, we have found a pair of stationary points with $\mathring{y}^{-,k}>\mathring{x}^{-,l}$. To obtain an according approximation to the bivariate dynamic integrals, we recall that a particular assumption leading to Eq. (3.33) was that the integrals are determined by the contributions from a close vicinity of the stationary points. Any contributions to a rapidly oscillating integral, formed far away from a stationary point was neglected. The bivariate parameter functions can then be expressed as

$$f_{ij}=\sum_{k,l}f_i^{\mathring{y}^{-,k}}\times f_j^{\mathring{x}^{-,l}}\,\Theta(\mathring{y}^{-,k}-\mathring{x}^{-,l}).$$

$$f_j^{\mathring{x}^{-,l}}=\int_{-\infty}^{\infty}dx^-\sum_{n_x=0}^{N_x}\frac{(x^--\mathring{x}^{-,l})^{n_x}}{n_x!}\left(\partial^{n_x}_{x^-}\psi^j_\mathcal{A}(x^-)\right)_{x^-=\mathring{x}^{-,l}}$$
$$\times\exp\left[-\mathrm{i}\sum_{m_x=0}^{M_x}\frac{(x^--\mathring{x}^{-,l})^m_x}{m_x!}\left(\partial^{m_x}_{x^-}S^{k_1}_{p_i,p_t}(x^-)\right)_{x^-=\Xi_0}\right]$$

$$f_i^{\mathring{y}^{-,k}}=\int_{-\infty}^{\infty}dy^-\sum_{n_y=0}^{N_y}\frac{(y^--\mathring{y}^{-,k})^{n_y}}{n_y!}\left(\partial^{n_y}_{y^-}\psi^j_\mathcal{A}(y^-)\right)_{y^-=\mathring{y}^{-,k}}$$

$$\times \exp\left[-\mathrm{i}\sum_{m_y=0}^{M_y}\frac{(y^- - \mathring{y}^{-,k})^{m_y}}{m_y!}\left(\partial_{y^-}^{m_y} S_{p_t,p_f}^{k_2}(y^-)\right)_{y^-=\Upsilon_0}\right] \tag{4.33}$$

In this expression we wrote the symbol $\Theta(\mathring{y}^{-,k} - \mathring{x}^{-,l})$ to denote that only stationary points with $\mathring{y}^{-,k} > \mathring{x}^{-,l}$ contribute to the final result. It is of particular importance this step function now enters Eq. (4.33) outside of the integrals over x^- and y^- which both cover the domain $(-\infty, \infty)$. The integrations over the two variables x^-, y^- thus formally decouple and the approximation for the bivariate dynamic integrals thus is a sum over all stationary points, which are found as solutions of Eq. (4.30). For each of the univariate dynamic integrals $f_i^{\mathring{y}^{-,k}}$, $f_j^{\mathring{x}^{-,l}}$ a saddle point approximation is already available in terms of Eqs. (3.36), (3.43), (3.47) and (3.48). In view of this decomposition then, in the relativistic regime, the complete phase ordered partial process decomposes in an analogous way according to

$$S_{fi}^{\text{p.o.}} \approx \sum_{l,k} S_{fi}^{(k)} \times S_{fi}^{(l)}\,\Theta\left(\mathring{y}^{-,k} - \mathring{x}^{-,l}\right), \tag{4.34}$$

where the $S_{fi}^{(l)}$ are the contributions from the lth stationary point $\mathring{c}^{-,l}$ to the NSCS scattering matrix element of the partial process at vertex c defined in Eq. (3.50). Equation (4.34) can be viewed as a discretized version of the general result Eq. (4.27). Discretized here has to be understood in the sense that the radiation of each separate NSCS event is the sum over the contributions from all stationary points. The sum in Eq. (4.34) then states, that to the phase ordered contribution of NDCS only those pairs of stationary points contribute, where the first photon is manifestly emitted before the second one. In all the previous arguments, of course, one has to pay attention that the overall amplitude of NDCS will always contain the cross term, depicted right in Fig. 4.1. We recall from Sect. 3.3 that the stationary points $\mathring{c}^{-,l}$ correspond to those phase instants η_c where the electron propagates into the direction of observation. The stationary points $\mathring{x}^{-,l}$ accordingly distinguish those points of the trajectory taken by a classical electron entering the laser field described by $A_L^\mu(\eta)$ with an initial momentum p_i^μ where it propagates into the direction (ϑ_1, φ_1). On the contrary the stationary points $\mathring{y}^{-,k}$ correspond to the propagation into the direction (ϑ_2, φ_2) of an electron entering the laser field with an initial momentum of p_t^μ. According to the discussion of Sect. 3.3 we then infer that the dynamics of the electron can be viewed as a combination of two classical trajectories, which are obtained from the different initial momenta p_i^μ and p_t^μ, respectively. The summation over all possible pairs of emission points indicates, that in this picture one has to sum all possible connection points of the two trajectories. These connection points are precisely the stationary points $\mathring{x}^{-,l}$ at which the first photon can be emitted. Since we saw that the overall process is dominated by the phase ordered contribution (compare Fig. 4.3), this property of the bivariate dynamic integrals approximately translates to the overall NDCS probability of Eq. (4.25). Of course, giving only the initial momenta of the two trajectories, they are only defined up to an additive integration constant namely the initial position of the electron. These two constants, however, are fixed by the in any case needed initial position of the electron before the scattering and by the physical requirement that the two trajectories need to be continuously joined. From the relations (4.32), we know the stationary points of the two emission events. We point out that, since $p_t\epsilon_L = (p_i - k_1)\epsilon_L$ and $p_t k_L = (p_i - k_1)k_L$, for $k_2^\mu = k_1^\mu$ (4.32b) exactly reproduces (4.32a), irrespective of p_t^μ. Since consequently both trajectories point into direction k_1 the at the same stationary points $\mathring{x}^{-,l}$, we conclude that also the derivatives of the two trajectories, joined at precisely these points, are continuous. This quasi-classical picture significantly simplifies the discussion of the results presented in this chapter and provides us with a deeper physical understanding of the physical process of two photon emission.

4.4 The Electron Propagator Dressed by an Arbitrary Plane Wave Field

As was mentioned in Chap. 1, a quantity of crucial importance in SF-QED computations, is the electron propagator dressed by an external plane wave laser field. In this chapter we are going to present a thorough analysis of the dressed electron propagator in an arbitrarily shaped external plane wave field. We will derive the propagator split up, given in Eq. (4.6), and employ it to prove that in a pulsed plane wave the propagator does not introduce divergences into the scattering matrix element. Furthermore we will comment on an ambiguity in the used split-up of the dressed electron propagator and the connected difficulty of assigning the physical terms *coherent* and *incoherent* to the photon emissions, introduced at the beginning of this chapter. As was mentioned there, this notion intends to distinguish between a pure quantum process and a classically realizable two photon emission. To be able to assign these interpretations to partial amplitudes of the scattering matrix element corresponding to Fig. 4.1, one would expect the partial amplitudes to satisfy a number of basic physical assumptions. Since in a quantum process both photons are emitted simultaneously, the coherent partial amplitude is expected to depend on only one single interaction point with the external field and to feature an off-shell momentum of the propagator electron. Since any process triggered by an external field, is more and more probable the longer the interaction with the external field lasts this quantum contribution should scale linearly in the pulse duration τ_L. A classical emission of two photons, on the other hand, is separable into two distinct emission events and should thus depend on two interaction points with the external field and scale quadratically in the laser pulse duration. And since a classical electron cannot change its mass, the propagator momentum, corresponding to the electron momentum between the two emissions, has to be on the mass-shell.

The technical discussion leading this section, will again be rather involved. A central result of it was already given in Eq. (4.6), and the remaining discussion is again summarized in a paragraph at the end of this section. We depart from Eq. (4.3), which is the expression of the dressed propagator with all trivial coordinate dependencies integrated out. The divergences in this equation can be isolated by means of the Sokhotski-Plemelj theorem [13]

$$\lim_{\epsilon\to 0}\frac{1}{p^2-m^2+i\epsilon}=\mathcal{P}\left(\frac{1}{p^2-m^2}\right)-\mathrm{i}\pi\,\delta(p^2-m^2), \tag{4.35}$$

where the symbol $\mathcal{P}(\dots)$ stands for the Cauchy principal value of the integrand enclosed in the brackets. Equation (4.3) then splits up into two contributions

$$\begin{aligned}G(y^-,x^-)=&\int\frac{dp^+}{2\pi}E_{p_t}(y^-)\mathcal{P}\left(\frac{p^+\gamma^-+\not{p}_t+m}{2p^+p_t^- -p_t^{\perp 2}-m^2}\right)\overline{E}_{p_t}(x^-)\mathrm{e}^{\mathrm{i}(p^+-p_t^+)(x^--y^-)}\\&-\frac{\mathrm{i}}{4p_t^-}E_{p_t}(y^-)\left(\not{p}_t+m\right)\overline{E}_{p_t}(x^-).\end{aligned} \tag{4.36}$$

In the Appendix E.1 now the following relation is proven

$$\int\limits_{-\infty}^{\infty}\mathcal{P}\left(\frac{\mathrm{e}^{\mathrm{i}px}}{p}\right)dp=\mathrm{i}\pi\,\mathrm{sgn}(x),$$

where $\mathrm{sgn}(x)$ is the sign function. Employing then variable transformation $\tilde{p}^+=p^+-p_t^+$, Eq. (4.36) turns into

$$\begin{aligned}G(y^-,x^-)&=G^{\mathrm{off}}(y^-,x^-)+G^{\mathrm{on}}(y^-,x^-)\\G^{\mathrm{off}}(y^-,x^-)&=E_{p_t}(y^-)\left(\frac{\not{k}_L}{2(p_tk_L)}\delta(y^--x^-)+\mathrm{i}\frac{\not{p}_t+m}{4p_t^-}\mathrm{sgn}(x^--y^-)\right)\overline{E}_{p_t}(x^-)\\G^{\mathrm{on}}(y^-,x^-)&=-\frac{\mathrm{i}}{4p_t^-}E_{p_t}(y^-)\left(\not{p}_t+m\right)\overline{E}_{p_t}(x^-).\end{aligned} \tag{4.37}$$

This way of splitting up the propagator manifestly distinguishes the two terms of the Sokhotski-Plemelj theorem (4.35). It is thus frequently interpreted as a separation into a term G^{on}, describing an intermediate electron on the mass shell ($p_t^2=m^2$), and G^{off}, with the intermediate electron being off the mass shell ($p_t\neq m^2$). We note that in the literature this particular way of splitting up the propagator is frequently used [4, 8, 12], in contrast to that position space decomposition, employed by us. To arrive at our result we have to reorganize Eq. (4.37) and make use of $\mathrm{sgn}(x^--y^-)-1=-2\Theta(y^--x^-)$, where $\Theta(y^--x^-)$ is the step function. It is then possible to find a propagator split up, separating between those contributions to the propagator yielding one and two dimensional integrations in the scattering matrix element, respectively. An equivalent view of this transformation is to attribute the imaginary contribution of the off-shell part to the on-shell part, yielding

$$G(y^-,x^-)=G^{\mathrm{p.o.}}(y^-,x^-)+G^{\mathrm{s.p.}}(y^-,x^-)$$

$$G^{\mathrm{p.o.}}(y^-, x^-) = -\mathrm{i}E_{p_t}(y^-)\frac{\not{p}_t + m}{2p_t^-}\overline{E}_{p_t}(x^-)\Theta(y^- - x^-)$$
$$G^{\mathrm{s.p.}}(y^-, x^-) = E_{p_t}(y^-)\frac{\not{k}_L}{2(p_t k_L)}\overline{E}_{p_t}(x^-)\delta(y^- - x^-). \qquad (4.38)$$

The superscripts *s.p.* and *p.o.* in Eq. (4.38) indicate the *same phase* and *phase ordered* contributions to the propagator, respectively. This notation is sensible, since in the same phase contribution it holds $\eta^x = k_L^+ x^- = k_L^+ y^- = \eta^y$, whereas the phase ordered contribution exhibits a succession of the two phase values according to $\eta^x < \eta^y$. In view of the above transformations we note that G^{on} contains terms with the manifest phase ordering $\eta^x > \eta^y$. The tempting interpretation that this reversed phase ordering describes a positronic contribution to the process depicted on the left hand side of Fig. 4.1, requiring a propagation backwards in time, is false. This can be concluded by noting that the propagator momentum component $p_t^- = p_i^- - k_1^- > 0$ is positive definite, whence a negative energy solution of Eq. (2.64a) cannot contribute to the on-shell contribution of Eq. (4.37). As mentioned in Sect. 4.2, we chose to use Eq. (4.38) for the computations presented in this work. Considering the regularization of the divergences that arise when considering a monochromatic plane wave as external laser field, we state that these are obviously absent when considering a pulsed laser field, as Eq. (4.23), which directly results from the employment of Eq. (4.38), does not contain any divergent quantities. How the divergences in the analysis of a monochromatic external laser wave explicitly reemerge, is demonstrated in section E.3. We can thus in fact attribute those divergences to the unphysical assumption of an infinitely extent external laser field. That the two separations, given in Eqs. (4.37) and (4.38), are not equal, is obvious as the two parts G^{off} and $G^{\mathrm{p.o.}}$, each containing the matrix contribution $(\not{p}_t + m)$, differ by a factor $(1 - \Theta(y^- - x^-))$. The decompositions of the diverging integrals f_{i0}, f_{0i}, as stated in Eqs. (4.21a) and (4.21b), result in a reverse attribution of parts of the bivariate dynamic integrals, arising from $G^{\mathrm{p.o.}}$ to the same phase contribution $G^{\mathrm{s.p.}}$. One might wonder if this operation might compensate the above described reattribution of propagator contributions, rendering the physically sensible split-up in position space equal to that one in momentum space. Since this renormalization of the dynamic integrals is only achieved for the scattering matrix element and not on a propagator level, it is necessary to investigate the decomposition of the full scattering matrix element, resulting from the propagator decomposition (4.37). Inserting this expression into Eq. (4.2) we find

$$S_{fi} = \mathcal{N}^{\mathrm{NDCS}}\left(M_{fi}^{\mathrm{off}} + M_{fi}^{\mathrm{on}}\right)\delta^{(3)}(p_i^{\perp} - k_1^{\perp} - k_2^{\perp} - p_f^{\perp})$$
$$M_{fi}^{\mathrm{off}} = \int dx^- dy^-\ \overline{\Psi}_{p_f}(y^-)\not{\epsilon}_2^* \mathrm{e}^{-\mathrm{i}k_2^+ y^-} E_{p_t}(y^-)$$
$$\times\left(\frac{\not{k}_L}{k_{L+}}\delta(y^- - x^-) + \mathrm{i}\frac{\not{p}_t + m}{2}\mathrm{sgn}(x^- - y^-)\right)$$
$$\times\,\overline{E}_{p_t}(x^-)\not{\epsilon}_1^* \mathrm{e}^{\mathrm{i}k_1^+ x^-}\Psi_{p_i}(x^-)$$

$$M_{fi}^{\text{on}} = -\mathrm{i}\int dx^- dy^- \overline{\Psi}_{p_f}(y^-)\not\epsilon_2^* \mathrm{e}^{-\mathrm{i}k_2^+ y^-} E_{p_t}(y^-)\left(\not p_t + m\right)$$
$$\times \overline{E}_{p_t}(x^-)\not\epsilon_1^* \mathrm{e}^{\mathrm{i}k_1^+ x^-} \Psi_{p_i}(x^-). \tag{4.39}$$

Comparing this result to Eq. (4.23) we find the partial amplitudes to be manifestly different. It is now the natural question which one of the decompositions is the natural one to facilitate an interpretation of the partial amplitudes as coherent and incoherent partial processes, respectively. In our opinion, it is physically sensible to consider the partial amplitudes of the scattering matrix element according to Eq. (4.23) instead of Eq. (4.39), as we give reasons for in the following. The first property which we wish to highlight is the time scaling of the partial processes resulting from Eq. (4.6). At a given space point the temporal duration of a laser pulse scales as $\tau_L \propto \Delta x^-$, whence we can equally analyze the scaling in the x^--interval over which the pulse is assumed to be nonzero. The scaling of the same phase and phase ordered partial processes, however, is easily inferred from the fact that $S_{fi}^{\text{s.p.}}$ involves one-dimensional integrals over the domain Δx^-, whereas the bivariate dynamic integrals entering $S_{fi}^{\text{p.o.}}$ are integrated over a two dimensional area of the dimension $\left(\Delta x^-\right)^2$. The expected time scaling of the coherent and incoherent partial processes thus is naturally reproduced by the same phase and phase ordered partial amplitudes, respectively. An even stronger argument for the validity of the propagator split up of Eq. (4.6) is the following. The incoherent process represents the succession of two separate NSCS events. The probability for each such process to occur, however, has been thoroughly studied in Chap. 3. If the two photon emissions are truly incoherent, the probability of the overall NDCS event must be expressible as a product of the probabilities of the two separate emissions, with the constraint that the emission of the second photon has to occur after the emission of the first one. This request is not to be confused with the result of Eq. (4.27), which is obtained on an amplitude level. For the previous conjecture to be realizable, however, the two processes must be formed in manifestly different space regions, for otherwise the separation into two successive processes becomes questionable. This request is always met in the relativistic regime $m\xi,\, \varepsilon_i \gg m$, where the formation length of the process is on the order of ξ^{-1} (see discussion in Sect. 3.3 and [14]). From Eq. (4.27) we know that, according to Eq. (3.57), in the relativistic regime the product of two NSCS probabilities, corresponding to an electron of initial momentum p_i^μ to emit a photon of momentum k_1^μ, thus changing its momentum to p_t^μ, and subsequently emitting a photon with momentum k_2^μ, changing its momentum to p_f^μ, is approximately given by a sum over all stationary points of the dynamic integrals, fulfilling the requested phase ordering

$$W_1^{\text{NSCS}} \times W_2^{\text{NSCS}} \overset{\xi \gg 1}{\approx} \left(\mathcal{N}_{\text{NSCS},1}^2 \mathcal{N}_{\text{NSCS},2}^2\right) \frac{1}{2} \sum_{\sigma_i,\sigma_t,\sigma_f} \int \frac{d\boldsymbol{k}_1 d\boldsymbol{k}_2 V^6}{(2\pi)^{18}}$$
$$\times \left(\sum_l \left|\bar{u}_{p_f,\sigma_f} M_{fi,2}^{(l)} u_{p_t,\sigma_t}\right|^2\right)\left(\sum_k \left|\bar{u}_{p_t,\sigma_t} M_{fi,1}^{(k)} u_{p_i,\sigma_i}\right|^2\right) \Theta\left(\mathring{y}^{-,k} - \mathring{x}^{-,l}\right), \tag{4.40}$$

where the spin indices of each electron are written explicitly and for notational simplicity we labeled the probabilities, the contribution to the reduced matrix elements from the respective stationary points and the normalization prefactors as W_1^{NSCS}, W_2^{NSCS}, $M_{fi,1}^{(k)}$, $M_{fi,2}^{(l)}$ and $\mathcal{N}_{\text{NSCS},1}$, $\mathcal{N}_{\text{NSCS},2}$ for the former and latter photon emission, respectively. The integration over the two final electron momenta has already been collapsed by virtue of the two corresponding energy momentum conserving δ-functions. The probability for the incoherent contribution to an NDCS amplitude from Sect. 4.3 is found to be given by a similar expression

$$W^{\text{NDCS}} = \left(\mathcal{N}_{\text{NSCS},1}^2 \mathcal{N}_{\text{NSCS},2}^2\right) \frac{1}{2} \sum_{\sigma_i,\sigma_f} \int \frac{d\boldsymbol{k}_1 d\boldsymbol{k}_2 V^6}{(2\pi)^{18}} \times \left| \sum_{k,l} \bar{u}_{p_f,\sigma_f} M_{fi,2}^{(l)} \left(\sum_{\sigma_t} u_{p_t,\sigma_t} \bar{u}_{p_t,\sigma_t} \right) M_{fi,1}^{(k)} u_{p_i,\sigma_i} \Theta\left(\mathring{y}^{-,k} - \mathring{x}^{-,l}\right) \right|^2 . \tag{4.41}$$

By virtue of the same argument employed in Sect. 3.3.1, that due to the strong oscillations of the transition probabilities in the final particles' phase space the interference terms between contributions from separate stationary points are suppressed, we can pull the sum $\sum_{k,l}$ out of the square. The only remaining difference between Eqs. (4.40) and (4.41) then is, that the sum over the spins of the transitional electron state is inside the modulus square in the latter expression. It is possible to rigorously evaluate the respective magnitude of the terms in the spin sum over equal and differing transitional electron spins by resorting to an explicit representation for the spinors u_{p_t,σ_t}, $\bar{u}_{p_t,\sigma_t}$ as e.g. given in Eq. (B.12b). In contrast to this tedious and uninstructive procedure we refer to the observation that spin interferences in NSCS are largely negligible [15]. We can thus also pull the sum over the transitional electron spins $\sum_{\sigma_t}$ out of the modulus square and arrive at an exact equivalent of Eq. (4.40). We have consequently argued that in the relativistic regime the incoherent NDCS probability can formally be written as the probability of two NSCS events happening one after the other

$$W^{\text{NDCS}} = \int_{-\infty}^{\infty} dx^- dy^- W_1^{\text{NSCS}}(x^-) \times W_2^{\text{NSCS}}(y^-) \Theta\left(y^- - x^-\right) . \tag{4.42}$$

Please note that this decomposition is fundamentally different from the property of independent emission spectra employed to describe the spectral structure in Fig. 4.3b, d. In Eq. (4.42) we explicitly include the possibility that the initial momentum of the second NSCS event p_t may differ from p_i due to recoil. We furthermore note that due to the differing energy momentum conserving δ-functions, the interference terms with the cross diagram from Fig. 4.1 vanish. We note that neither of the partial amplitudes in Eq. (4.39) features a corresponding ordering of the respective phases, whence it is impossible to derive an expression corresponding to Eq. (4.42) from it.

We are obliged to state that despite its advantages in terms of physical interpretability, the presented split-up prescription is not flawless. In [4, 8, 12] it was argued that the proper split-up for the two separate contributions to the electron propagator should feature respectively differing energy-momentum conservation laws. In fact it was shown that for pair-creation in a strong laser wave the incoherent contribution has to fulfill separate energy momentum conservation laws for each of the two separate processes. In the here studied case of NDCS we would accordingly expect the incoherent contribution to feature two energy momentum conservation laws of the form $\delta^{(4)}(p_i + sk_L - k_1 - p)\delta^{(4)}(p + rk_L - k_2 - p_f)\delta(p^2 - m^2)$, where s, r can be viewed as a continuous generalization of the photon numbers, found in monochromatic analyses. The incoherent contribution, on the other hand, only has to fulfill the single overall energy-momentum conservation $\delta^{(3)}(p_i^{\perp} - k_1^{\perp} - k_2^{\perp} - p_f^{\perp})$. This latter request is equivalent to the overall energy momentum conservation of Eq. (4.23). It thus suffices to analyze only the bivariate dynamic integrals. It is easily seen that they in fact reproduce the two perpendicular conservation laws, by restoring the integration over the p^--component of the propagator momentum in Eq. (4.1). To formally recover an on-shell condition, we have to employ the integral representation of the step function Eq. (E.2), as well as the continuous Fourier transform of the univariate dynamic integrals,

$$\psi^i_{\mathcal{A}}(x^-)\mathrm{e}^{-\mathrm{i}S^k_{p,q}(x^-)} = \int ds\; C^{k,i}_{p,q}(s)\mathrm{e}^{-\mathrm{i}(p^+ + sk_L^+ - k^+ - q^+)x^-}, \tag{4.43}$$

where the coefficients are defined according to [14]

$$C^{k,i}_{p,q}(s) = \frac{k_L^+}{2\pi}\int dc^-\,\psi^i_{\mathcal{A}}(c^-)\mathrm{e}^{-\mathrm{i}(g^k_{p,q}(c^-) - sk_L^+ c^-)}, \tag{4.44}$$

with $g^k_{p,q}(c^-)$ being the integral part of the action $S^k_{p,q}(c^-)$, as defined in Eq. (4.14). We note that in the monochromatic limit the integral definition in Eq. (4.44) will again collapse to a sum over photon orders (see Sect. 2.2.3). Applying the Sokhotski-Plemelj theorem Eq. (4.35) to the denominator of Eq. (E.2) and retracting the variable change $p^+ = \tilde{p}^+ + p_t^+$, introduced for evaluating Eq. (4.36), we can express the bivariate dynamic integrals as

$$\begin{aligned} f_{ij} = &- \mathrm{i}(2\pi)\int dp^+ ds dr\; C^{k_1,j}_{p_i,p}(s)C^{k_2,i}_{p,p_f}(l)\delta\left(p_i^+ + sk_L^+ - k_1^+ - p^+\right) \\ &\times \delta\left(p^+ + rk_L^+ - k_2^+ - p_f^+\right)\left[\mathcal{P}\left(\frac{1}{p^+ - p_t^+}\right) - \mathrm{i}\pi\delta(p^+ - p_t^+)\right]. \end{aligned} \tag{4.45}$$

This expression obviously does not reduce to an expression featuring solely two separate δ-functions with the intermediate electron fulfilling an on-shell condition

$$f_{ij} \nrightarrow \delta\left(p_i^+ + sk_L^+ - k_1^+ - p^+\right)\delta\left(p^+ + rk_L^+ - k_2^+ - p^+\right)\delta(p^+ - p_t^+),$$

which in combination with the perpendicular energy momentum conservation of Eq. (4.23) would yield the correct two step energy momentum conservation. We thus find the cumbersome result that the dressed electron propagator can be either understood as separable into an on- and off-shell part, or a same phase and a phase ordered contribution. It is, however, not possible to present a rigorous decomposition of the dressed propagator satisfying all requirements to interpret the partial processes as physically coherent and incoherent, respectively. The physical picture, that an electron in a laser field can become on-shell and propagate over a finite distance between two separate physical interactions, rendering the emission of the two light quanta incoherent, whereas an electron with off-shell momentum interacts multiply at the very same space-time point, rendering its interaction coherent, is clearly flawed. This result translates to any higher order SF-QED process requiring the use of dressed propagators. It is, however, only seemingly contradictory. The mentioned physical picture relies intrinsically on the assumption of a localizable electron. In fact, if the electron has a large position uncertainty, the notion of interaction at the same or different space-time points loses its validity. We can thus only hope to unambiguously identify coherent and incoherent portions of the scattering matrix element in the case that the electron can be viewed as quasi-classical. As we already took advantage of in the derivation of Eq. (4.42), this is the case in the ultra-relativistic regime. As was argued in Sect. 3.3, in this regime the coherence interval of photon emission by an electron scales as ξ^{-1}. Since in this regime consequently the notion of a localized electron is approximately valid, we are going to investigate the electron propagator in the limit $m\xi, \varepsilon_i \gg m$, to recover a clear physical image of coherent and incoherent two photon emission. We saw that employing the propagator split up Eq. (4.6), one cannot recover the energy momentum conservation of a two step process, as is captured in Eq. (4.45). The disturbing contribution is that of the principal value integration. In the relativistic regime, however, this contribution evaluates to

$$\int dp^+ ds\delta\left(p_i^+ + sk_L^+ - k_1^+ - p^+\right)\mathcal{P}\left(\frac{1}{p^+ - p_t^+}\right)$$
$$= \int ds\, \mathcal{P}\left(\frac{1}{k_L^+(s + \gamma_V^x)}\right). \tag{4.46}$$

But according to the discussion of Sects. 3.3 and 4.3 it holds $\gamma_V^x \sim \xi^3$, whence we conclude that the above contribution to the bivariate dynamic integrals is strongly suppressed in the relativistic regime with respect term proportional to $\delta(p^+ - p_t^+)$ in Eq. (4.45). If we accordingly neglect the principal value integration in that equation, however, we find that the propagator split up employed in Eq. (4.6) satisfies all of the previously mentioned requirements for a physical interpretation of the partial amplitudes.

4.4.1 Summarizing the Above Discussion

We have seen how the electron propagator dressed by an arbitrary plane wave field naturally splits up into two parts. Employing this split up propagator we saw, that the divergences arising in the analysis of monochromatic external laser waves [5, 6, 11], were due to the unphysical assumption of a monochromatic plane wave. We have further unveiled the impossibility of generally interpreting the partial amplitudes of any higher order SF-QED process as coherent or incoherent in a physical sense, due to the delocalization of the involved massive particles. On the contrary we saw, that in the relativistic regime, where the classical picture of localized particle trajectories regains physical validity, it is possible to meaningfully identify a coherent and incoherent contribution to the two-photon emission probability. In this respect it is good news that the split-up Eq. (4.6) fulfills all physically intuitive requirements to interpret the phase ordered and the same phase contributions to Eq. (4.23) as the incoherent and coherent physical partial processes, respectively. We conjectured that the presented results are generalizable to arbitrary processes involving the dressed electron propagator.

4.5 Photon Correlation Due to Quantum Effects

In this section we wish to sketch a particular application of the stationary phase approximation presented in the previous section. We are going to make use of the there introduced simple picture of the electron's dynamics inside the laser field in terms of two continuously joined trajectories. Secondly we demonstrate that the corresponding understanding of these dynamics lays path for an unbiased detection of an NDCS signal. In view of the detectability of a NDCS signal it was pointed out that the integrated two-photon emission probability is mostly exceeded by the single photon emission probability [4]. The latter quantity was thoroughly studied in the previous Chap. 3. In particular we could show that the angular distribution of the NSCS signal is confined to an emission cone, which in the case $\varepsilon_i \gg m\xi$ is always confined to emission angles deviating from π by no more than $\vartheta_0 = {}^{m\xi}/_{\varepsilon_i}$. Any photon emission outside of this cone then clearly originates from a process other than single photon emission and can thus be discriminated from it. We are going to demonstrate that in fact, the NDCS analysis of the previous section predicts photon emission to directions outside of this NSCS emission cone. In simple terms the reason for this phenomenon is the recoil a highly energetic photon exerts on the electron. If recoil is significant, the initial momentum p_t^μ of the second classical trajectory significantly differs from p_i^μ. On the accordingly changed trajectory an electron will propagate and thus radiate towards different directions as on the unchanged trajectory with initial momentum p_i^μ. To demonstrate the above claim we are going to compare NDCS spectra, obtained for the scattering of an electron from a laser pulse modeled by Eq. (2.6) with $n_C = 2$ and observed in a coordinate frame according to Fig. 2.1, where we choose the observation direction of one photon to always lie inside the

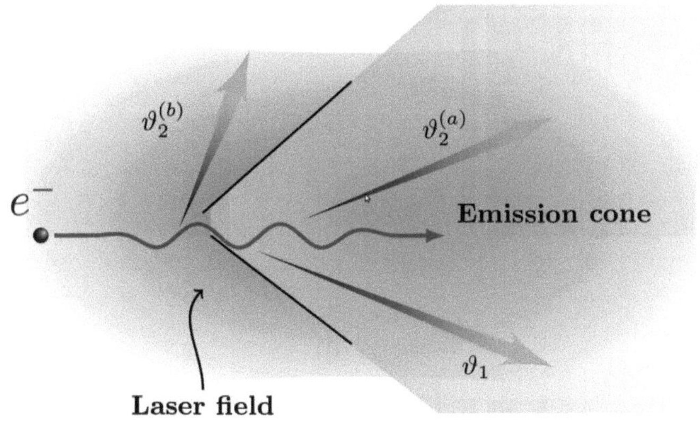

Fig. 4.5 Schematic picture of the chosen observation directions $\vartheta_2^{(a)}$ $\left(\vartheta_2^{(b)}\right)$ lying inside (outside) the electron's NSCS emission cone. For clarity the case of $\varphi_2 = \varphi_1\left(\vartheta_2^{(a,b)}\right.$ mirrored at the axis of the electron's initial propagation direction.$\left.\right)$ is not shown

NSCS emission cone $|\pi - \vartheta_1| < \vartheta_0, \varphi_1 = 0$. For the other photon, however, we distinguish the four cases $\left|\pi - \vartheta_2^{(a)}\right| < \vartheta_0$ and $\left|\pi - \vartheta_2^{(b)}\right| > \vartheta_0$ and $\varphi_2 = 0, \pi$ (see Fig. 4.5). The fixing of the two azimuthal observation angles φ_1, φ_2 to the plane of laser polarization is in accordance with the arguments given in Sect. 3.3. As choices for the polar observation angles we employ $\vartheta_1 = \vartheta_2^{(a)} = \pi - \vartheta_0/2$ as well as an observation direction slightly outside of the NSCS emission cone $\vartheta_2^{(b)} = \pi - 1.1\vartheta_0$. According to the discussion of Sect. 4.2, we will show the total energy emitted, which is dominated by the phase ordered contribution. We are furthermore going to show the energy emission spectra with a logarithmic $\omega_{1,2}$-scaling, since we expect a significant contribution to the energy spectra in the regimes $\omega_{1,2} \ll \omega_{2,1}$. This expectation is based on the observation that with significant recoil it will hold $\varepsilon_t < \varepsilon_i$ and hence, due to the cubic dependence of the emitted photon frequencies on the electron's energy, the emitted photon frequencies will strongly differ. To quantify the effect of photon recoil we begin the analysis in the parameter regime $\chi \ll 1$, where recoil is expected to be negligible. To study this regime we consider an electron of initial energy $\varepsilon_i = 50$ MeV being scattered from a laser pulse of peak intensity $I_L = 5 \times 10^{20}$ W/cm^2, corresponding to $\xi \approx 15$. For these parameters we find $\chi \approx 10^{-2} \ll 1$, whence we conclude that this case probes the desired regime. In Fig. 4.6 we show the energy emission spectra $dE/(\prod_{i=1,2} d\omega_i d\Omega_i)$ obtained from the numerical evaluation of Eq. (4.25). Comparing the spectra, predicted for an observation of both photons inside the emission cone (Fig. 4.6a, c) to Fig. 4.3d, we recover the known results of both photons being emitted at the same energy scales $\omega_1 \sim \omega_2$. These energy emission spectra closely resemble the separable structure found in Sect. 4.2 as the product of two independent NSCS emission probabilities for each photon. On the other hand, it is apparent that outside of this cone there is no

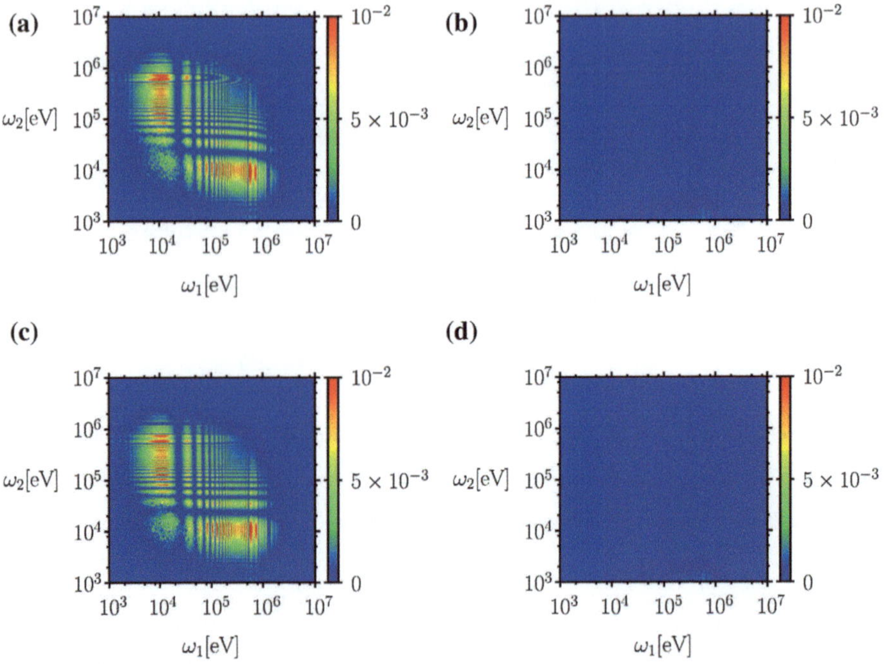

Fig. 4.6 Energy emission spectra $dE/\left(\prod_{i=1,2} d\omega_i d\Omega_i\right)$ for $\varepsilon_i = 50\,\text{MeV}$, $I_L = 5 \times 10^{20}\ \text{W/cm}^2$, observed at the specified chosen observation directions (ϑ_1, φ_1) and $\vartheta_2^{(a)}$ $\left(\vartheta_2^{(b)}\right)$ lying inside (outside) the electron's NSCS emission cone and $\varphi_2 = 0, \pi$. The spectra are given in [eV^{-1} sr^{-2}]. **a** Observation of both photons inside the NSCS emission cone ($\vartheta_2^{(a)}$) at $\varphi_2 = \pi \neq \varphi_1$. **b** Observation of one photon outside of the NSCS emission cone ($\vartheta_2^{(b)}$) at $\varphi_2 = \pi \neq \varphi_1$. **c** Observation of both photons inside the NSCS emission cone ($\vartheta_2^{(a)}$) at $\varphi_2 = \varphi_1 = 0$. **d** Observation of one photon outside of the NSCS emission cone ($\vartheta_2^{(b)}$) at $\varphi_2 = \varphi_1 = 0$

photon emission observed (Fig. 4.6b, d). We furthermore observe an almost perfect symmetry between the spectra shown for $\varphi_2 = 0$ and $\varphi_2 = \pi$, respectively. All these features can be properly explained by means of the physical picture of smoothly joined trajectories, established in Sect. 4.3. In the case $\chi \ll 1$ the recoil of the emitted photon on the electron is negligible and the electron's momentum after each individual emission event is approximately equal to its momentum before the photon emission. In the case studied in Fig. 4.6 we can thus approximate $p_f \approx p_t \approx p_i$ and the two separate electron trajectories coincide. In Fig. 4.7 we show the trajectory of an electron with initial momentum p_i, where the points, at which emission into $\boldsymbol{n}_1$ is feasible (the stationary points $\mathring{x}^{-,l}$), are denoted by black crosses. Along the trajectory the classical electron energy is color coded according to the scale on the left, where $\varepsilon_{\max}$ is the maximum energy the electron attains inside the laser field. In addition we show the trajectory of an electron with initial momentum p_t, resulting from p_i after the emission of a photon with frequency $\omega_1 = 2 \times 10^4$ eV

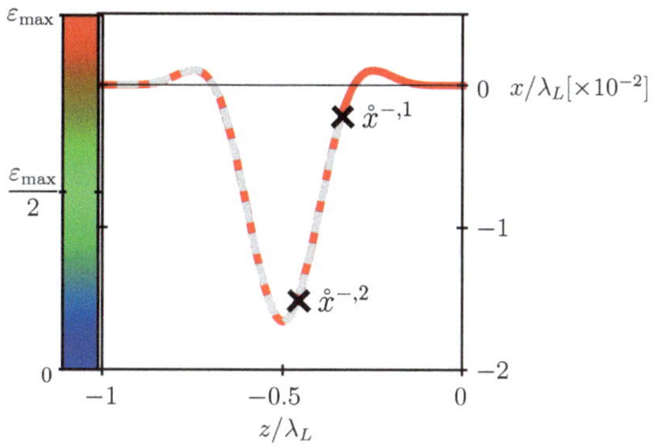

Fig. 4.7 Trajectory of an electron with initial energy $\varepsilon_i = 50\,\text{MeV}$ (*solid*) smoothly joined at $\mathring{x}_1^-$ to the trajectory of an electron with initial momentum p_t (*gray dashes*). The *color code* gives the classical energy of the electron inside the laser field

(corresponding to the maximum of the emission spectrum, analogous to Fig. 4.6) along $\boldsymbol{n}_1$, continuously joined to the former trajectory at $\mathring{x}^{-,1}$ in gray dashes. It is apparent that the two trajectories are indistinguishable. The color coding for the second trajectory had to be omitted, since red dashes are not visible on top of the first trajectory. Nevertheless we note that the difference between the electron energies on both trajectories is negligible. The absence of an NDCS signal outside of the NSCS emission cone (see Fig. 4.6b, d) is then explained analogously to the discussion of Sect. 3.4. If the electron trajectory is unchanged by photon emission, then throughout the whole scattering it does not propagate towards directions pointing outside of the NSCS emission cone. We call attention to the fact, that the previous discussion may not be understood in such a way as to allow a distinction which of the two photons is emitted before or after the respectively other one. One has to keep in mind that the above discussion solely referred to the left diagram of Fig. 4.1. In the analysis of the overall scattering process, of course the cross term, represented by the right diagram of the previous figure, has to be included. Each of these partial diagrams, however, indeed does allow for a definite discrimination between the photon emitted first and second.

The situation now changes drastically, when electron recoil becomes significant. In order to study the impact of photon recoil, but keeping experimental feasibility not outside of our considerations, we study the case of an electron with initial energy $\varepsilon_i = 2.5$ GeV scattered from a laser pulse of peak intensity $I_L = 3 \times 10^{21}$ W/cm^2, corresponding to $\xi \approx 37$, modeled by the same shape function and observed at the same angles as in the former case. In Fig. 4.8 we show corresponding energy emission spectra. In fact, as suspected, the spectra qualitatively notably differ from Fig. 4.6 in four aspects. First we find that the sum of the emitted photon frequencies approaches the quantum mechanical cutoff energy, as obtained from Eq. (4.5), which hints at the importance of quantum effects. Second we note that inside of the emission cone (Fig. 4.8a) instead of both photons being emitted at a comparable energy, the spectra are strongly peaked at the regions $\omega_1 \gg \omega_2$ and $\omega_2 \gg \omega_1$. Thirdly we observe that the symmetry between the cases $\varphi_2 = \varphi_1$ and $\varphi_2 \neq \varphi_1$ is destroyed. And finally, most strikingly, we find that unlike in the regime $\chi \ll 1$ there is emission predicted

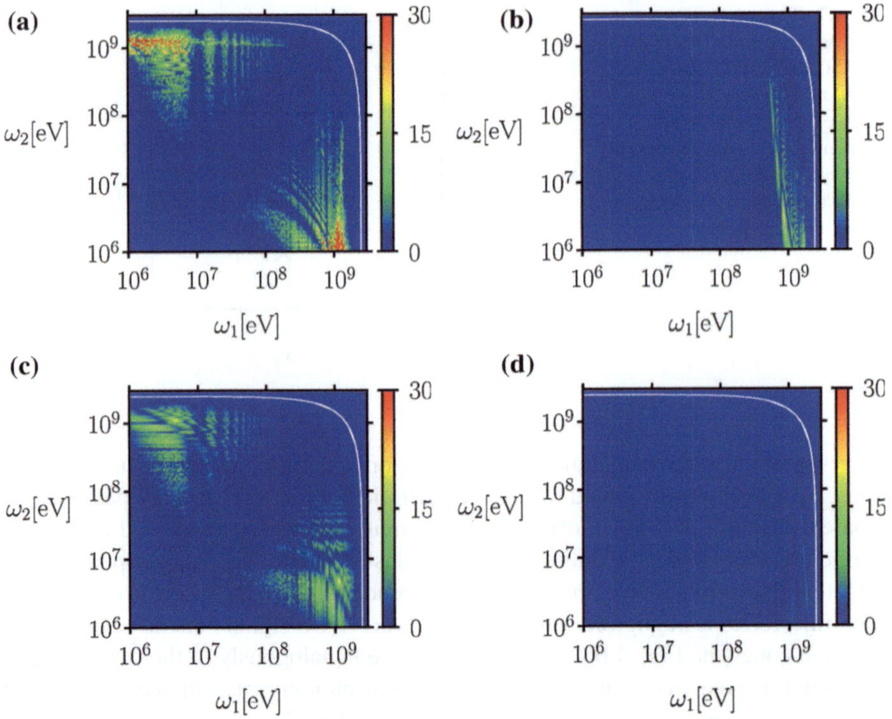

Fig. 4.8 Energy emission spectra $dE/\left(\prod_{i=1,2} d\omega_i d\Omega_i\right)$ for $\varepsilon_i = 2.5\,\mathrm{GeV}$, $I_L = 3\times10^{21}\ \mathrm{W/cm^2}$, observed at the specified chosen observation directions (ϑ_1, φ_1) and $\vartheta_2^{(a)}$ $\left(\vartheta_2^{(b)}\right)$ lying inside (outside) the electron's NSCS emission cone and $\varphi_2 = 0, \pi$. The *white lines* indicate the quantum mechanical cutoff energy, according to Eq. (4.5). The spectra are given in $[\mathrm{eV}^{-1}\ \mathrm{sr}^{-2}]$. **a** Observation of both photons inside the NSCS emission cone $(\vartheta_2^{(a)})$ at $\varphi_2 = \pi \neq \varphi_1$. **b** Observation of one photon outside of the NSCS emission cone $(\vartheta_2^{(b)})$ at $\varphi_2 = \pi \neq \varphi_1$. **c** Observation of both photons inside the NSCS emission cone $(\vartheta_2^{(a)})$ at $\varphi_2 = \varphi_1 = 0$. **d** Observation of one photon outside of the NSCS emission cone $(\vartheta_2^{(b)})$ at $\varphi_2 = \varphi_1 = 0$

outside of the NSCS emission cone (see Fig. 4.8b, d, albeit in the latter case the emission is less pronounced than in the former). The previously established picture of two continuously joined trajectories now enables us to consistently explain all these observations in a remarkably simple way. We note, that in the regime $\chi \sim 1$ the transitional momentum p_t^μ is manifestly different from p_i^μ. To obtain a particular numerical value for p_t^μ, we consider the emission of a photon with energy $\omega_1 = 8\times10^8$ eV (corresponding to the threshold above which there is significant emission predicted in Fig. 4.8b) along $\boldsymbol{n}_1$. The two resulting trajectories, taken by an electron entering the laser pulse with these two respective momenta, are shown in Fig. 4.9. In this figure we chose to join the two trajectories at $\mathring{x}^{-,1}$, the first possible emission point of the photon with wave vector k_1. The colored trajectory segments indicate the actual trajectory the electron takes, when the two single trajectories are joined at

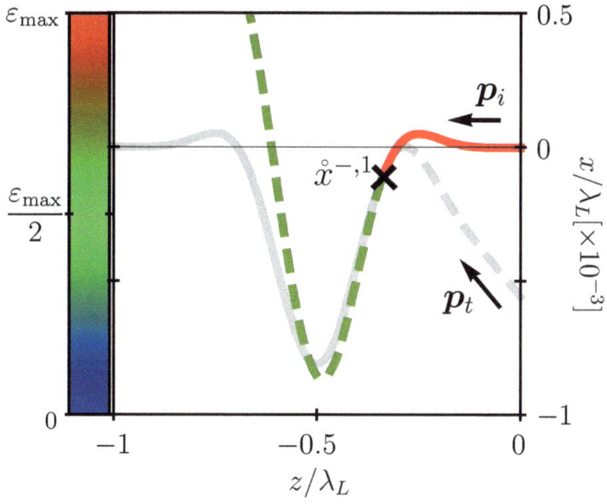

Fig. 4.9 Trajectory of an electron with initial energy $\varepsilon_i = 2.5\,\text{GeV}$ (*solid*) smoothly joined at $\mathring{x}_1^-$ to the trajectory of an electron with initial momentum p_t (*dashed*). The *color code* gives the classical energy of the electron inside the laser field

this point. The analysis of the case where the two trajectories are joined at the second stationary point $\mathring{x}^{-,2}$ is analogous and equally enters the final result. The color code gives the classical electron energy along each trajectory segment, analogous to Fig. 4.7. We read off that the energy of the electron discontinuously decreases significantly at the emission point, where the trajectories otherwise are continuously joined. We specifically note that the derivatives of the two trajectories coincide at the junction as was expected based on Eq. (4.31). The discontinuity in the energy is a result of the photon carrying away a non-negligible portion of the electron's energy and a clear quantum effect. Due to the cubic scaling of the emitted photon energies in the electron's instantaneous energy (see Eq. (2.23)) any photon emitted by the electron after it emitted the first one, and hence follows the trajectory to the initial momentum p_t^μ, will feature a significantly smaller energy than the first emitted photon. The strong asymmetry in the emitted photon energies, observed in Fig. 4.8a, c, is consequently explained as a signature of the energy loss of the electron upon emission of a high energetic photon. In addition to the energy asymmetry we can also explain the loss of symmetry between the cases $\varphi_2 = 0$ and $\varphi_2 = \pi$, based on the classical trajectory picture of Fig. 4.9. We note that the trajectory to the initial momentum p_t^μ does not point to equal ϑ-ranges for $\varphi = 0$ and $\varphi = \pi$, respectively. This behaviour is explained by the fact that the strength of the deflection of an electron with initial energy ε by the laser field with the invariant intensity parameter ξ is determined by the ratio $m\xi/\varepsilon$, which is changed between the two emission events if the first photon exerts a recoil on the electron. We thus cannot expect the energy emission spectra in these two azimuthal angle regimes to be similar. By an analogous argument we can also provide an explanation for the prediction of radiation outside of the NSCS emission cone. Analogous to the discussion of Sect. 3.4 we can quantify the polar angle range, into which the second photon can be emitted, by means of Eq. (4.32b). Expanding this expression in the middle of that equation and using the

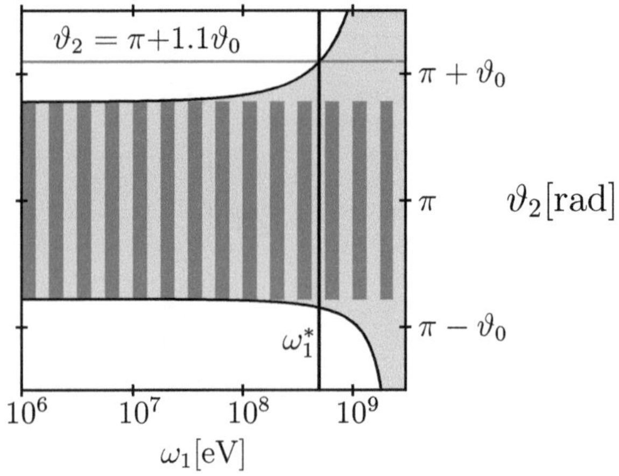

Fig. 4.10 Angular range where the emission of the second photon in a scenario specified to obtain Fig. 4.8 is possible. The angular range into which the second photon with wave vector k_2^μ can be emitted, is *light gray* and the NSCS emission cone is indicated by *darker vertical stripes*

approximations $\varepsilon_i \gg 1$, $\pi - \vartheta_2 \ll 1$ we find the stationary points for the second photon emission to be determined by the condition

$$\psi_{\mathcal{A}}(\mathring{y}^{-,k}) = -\frac{\alpha_V^y}{2\beta_V^y} = \frac{1}{\varepsilon_i \vartheta_0}\left(\cos(\varphi_2)\cot\left(\frac{\vartheta_2}{2}\right)(p_i n_L) - \cos(\varphi_2)\cot\left(\frac{\vartheta_2}{2}\right)(k_1 n_L) + k_1^x\right)$$
$$\approx \cos(\varphi_2)\Delta\theta_2 + \left(\frac{\omega_1}{\varepsilon_i}\right)\left(\frac{\cos(\varphi_1)}{2} - \cos(\varphi_2)\Delta\theta_2\right), \tag{4.47}$$

where we defined the quantity $\Delta\theta_2 = {}^{(\pi-\vartheta_2)}/_{\vartheta_0}$. In Eq. (4.47) the term proportional to ${}^{\omega_1}/_{\varepsilon_i}$ represents the influence of the recoil exerted by the first emitted photon and one can clearly trace the increasing importance of this term for increasing ω_1. Fixing now the observation angles (ϑ_1, φ_1) to the values chosen for obtaining Fig. 4.8, the above expression turns into a boundary condition for (ϑ_2, φ_2) as a function of ω_1. In analogy to the discussion of Sect. 3.4 we map $\varphi_2 = 0$ to $\vartheta_2 < \pi$ and $\varphi_2 = \pi$ to $\vartheta_2 > \pi$. In Fig. 4.10 we show the resulting emission cone of the second photon's emission as a function ω_1. As an orientation also the NSCS emission cone is indicated by darker vertical stripes.

In this figure it clearly shows how for increasing ω_1 the emission cone gets more and more spread out, rendering the emission of the second photon to directions outside of the NSCS emission cone feasible. The black vertical line, labeled ω_1^*, gives the frequency threshold, above which photon emission into $(\vartheta_2^{(b)}, \varphi_2 = \pi)$ becomes feasible. We also read off that the emission into $(\vartheta_2^{(b)}, \varphi_2 = 0)$ becomes significant only beyond considerably larger values of ω_1 (compare Fig. 4.8d), whence the discussion of photon emission outside of the NSCS emission cone is legitimately limited to the former azimuthal angle region. We point out, that in Fig. 4.8b this threshold ω_1^* is well reproduced. The physical explanation for the just described effect in terms of the two trajectory picture is the following: As we saw in Fig. 4.9

a decreased electron energy implies a stronger bending of the electron trajectory in the laser field. The larger ω_1 now grows, the smaller the electron's energy after the emission of the first photon accordingly is, rendering its trajectory more strongly bent. A stronger bending of the trajectory, however, leads the emission of the second photon into an increasingly large angular range. Summing up numerically the emission probability, corresponding to the energy emission spectra given in Fig. 4.8a, b, we conclude that the integrated probability of photon emission outside of the NSCS emission cone, albeit smaller than inside the cone, still is on the order of 0.1 for the considered experimental parameters. This should be well detectable, having in mind that typical electron beams contain on the order of 10^8 electrons per shot.

References

1. P.G. Thirolf, D. Habs, A. Henig, D. Jung, D. Kiefer, C. Lang, J. Schreiber, C. Maia, G. Schaller, R. Schützhold, T. Tajima, Signatures of the Unruh effect via high-power, short-pulse lasers. Eur. Phys. J. D **55**, 379 (2009)
2. R.J. Glauber, Some notes on multiple-boson processes. Phys. Rev. **84**, 395 (1951)
3. D. Melrose, A classical counterpart to double Compton scattering. Nuovo Cim. A **7**, 669 (1972)
4. D. Seipt, B. Kämpfer, Two-photon Compton process in pulsed intense laser fields. Phys. Rev. D **85**, 101701 (2012)
5. E. Lötstedt, U.D. Jentschura, Nonperturbative treatment of double Compton backscattering in intense laser fields. Phys. Rev. Lett. **103**, 110404 (2009)
6. E. Lötstedt, U.D. Jentschura, Correlated two-photon emission by transitions of Dirac-Volkov states in intense laser fields: QED predictions. Phys. Rev. A **80**, 053419 (2009)
7. F. Mackenroth, A. Di Piazza, Nonlinear double Compton scattering in the full quantum regime. Phys. Rev. Lett. **110**, 070402 (2013)
8. A. Ilderton, Trident pair production in strong laser pulses. Phys. Rev. Lett. **106**, 020404 (2011)
9. F. Mandl, T.H.R. Skyrme, The theory of the double Compton effect. P. Roy. Soc. Lond. A Mat. **215**, 497 (1952)
10. P.E. Cavanagh, The double Compton effect. Phys. Rev. **87**, 1131 (1952)
11. V.P. Oleinik, Resonance effects in field of an intense laser beam. Sov. Phys. JETP **25**, 697 (1967)
12. H. Hu, C. Müller, C.H. Keitel, Complete QED theory of multiphoton trident pair production in strong laser fields. Phys. Rev. Lett. **105**, 080401 (2010)
13. F.W. Olver, D.W. Lozier, R.F. Boisvert, C.W. Clark, *NIST Handbook of Mathematical Functions* (Cambridge University Press, New York, 2010)
14. V.I. Ritus, Quantum effects of the interaction of elementary particles with an intense electromagnetic field. J. Russ. Laser Res. **6**, 497 (1985)
15. M. Boca, V. Dinu, V. Florescu, Spin effects in nonlinear Compton scattering in a plane-wave laser pulse. Nucl. Instrum. Meth. B **279**, 12 (2012)

Chapter 5
Numerical Integration of Strongly Oscillating Functions

The purpose of computing is insight, not numbers.
Richard Hamming

The numerical evaluation of the dynamic integrals f_i, f_{ij}, given by Eqs. (3.5), (4.9) and (4.13) can be tedious, due to the large values the exponential phase $S^{k_1}_{p_i,p_f}(x^-)$ may grow to, for the emission of a photon with wave vector k_1^μ by an electron changing its momentum from p_i^μ to p_f^μ. The scaling of this exponential phase of the dynamic integrals is given by the parameters α_V, β_V and γ_V, which in Sect. 3.3 were shown to scale as $\alpha_V, \beta_V, \gamma_V \sim \xi^3$ in the parameter regime $\xi, \gamma_i \gg 1$. Thus the action can become very large rendering the integral functions strongly oscillating.

5.1 Filon Type Integration

The purpose of numerical quadrature routines is to find numerical values for integrals of the type

$$\mathbb{I}(a,b) = \int_{-b}^{a} dx\, \varphi(x)\, \mathrm{e}^{\mathrm{i}\omega x}, \tag{5.1}$$

with a constant oscillation frequency ω. In the case of an improper integration, i.e. $a, b = \infty$, we note that unless the integral can be evaluated analytically, one has to introduce a finite upper and lower boundary of the integration domain, which is suited to capture the essence of the integral. In this respect we note that for the dynamic integrals defined in Eqs. (3.5), (4.9) and (4.13), which we look for a numerical estimate for, this implies the form

K. F. Mackenroth, *Quantum Radiation in Ultra-Intense Laser Pulses*, Springer Theses, DOI: 10.1007/978-3-319-07740-6_5,

$$f_i \approx \int_{x^-_{\text{min}}}^{x^-_{\text{max}}} dx^- \psi^i_{\mathcal{A}}(x^-) \mathrm{e}^{-\mathrm{i}\, S^{k_1}_{p_i, p_f}(x^-)} \tag{5.2a}$$

$$f_{ij} \approx \int_{x^-_{\text{min}}}^{x^-_{\text{max}}} dx^- dy^- \psi^j_{\mathcal{A}}(x^-) \psi^i_{\mathcal{A}}(y^-) \mathrm{e}^{-\mathrm{i}\left(S^{k_1}_{p_i, p_t}(x^-) + S^{k_2}_{p_t, p_f}(y^-)\right)}. \tag{5.2b}$$

This approximate form, however, is either exactly obtainable in case the shape function $\psi^i_{\mathcal{A}}(x^-)$ is defined only on a finite interval $x^- \in [x^-_{\text{min}}, x^-_{\text{max}}]$, as in Eqs. (2.6) and (2.8). Or it at least represents a very good approximation for shape functions strongly damped outside of a finite interval, as is natural for a laser field of finite temporal duration. For the following general discussion, however, we will consider integrals of the form of Eq. (5.1) and later return to the special case of Eqs. (5.2a) and (5.2b). The problem that the oscillation of the integrand in Eqs. (5.2a) and (5.2b) does not feature a constant frequency, will also be resolved later. Furthermore we will restrict the general discussion in this chapter to the univariate case. The generalization to several integration dimensions is straightforward with some intricacies discussed in Sect. 5.3.

Usual univariate numerical quadrature routines resort to interpolating the integrand by a suitable function which can be analytically integrated and provides a suitable approximation for the integral itself. As discussed above, in the case of interest here, the oscillation of the integrand is assumed to be very rapid. Conventional schemes thus approximate the integrand only poorly. It is then customary to break down the integration interval into N subintervals and apply the approximation method on each interval separately. The result will be a sum of function values of the integrand each weighted with an appropriate coefficient, to be determined by the specific approximation for the integrand.

$$\mathbb{A} = \sum_{n=1}^{N} w_n \,\varphi_n, \tag{5.3}$$

with the position of the sampling nodes $x_n = -b + nh$, where $h = {}^{(a+b)}/_{(N-1)}$ is the step width, the function values at these nodes $\varphi_n = \varphi(x_n)$ and the weighting coefficients w_n. Intuitively one would expect that to approximate an oscillatory integral as Eq. (5.1) one would have to break up the integration interval into more and more subintervals the stronger the integrand oscillates. This would finally result in computationally untreatably many integration steps. Other than this obvious argument it was also shown that the error for a fixed number of interpolation points increases with increasing oscillation frequency [1]. This problem can be resolved by resorting to a quadrature scheme first proposed by Filon in 1928 [2] and frequently reinvestigated [1] for integrals of the type (5.1), where $\varphi(x)$ is assumed to be a slowly oscillating preexponential function. Filon suggested to split up the integration interval into N subintervals and to apply a three point quadrature scheme on each interval, to interpolate not the whole integrand but only the preexponential function $\varphi(x)$. The whole integral will then be approximated at $2n+1$ nodes and it was shown that the error derived from such an interpolation scheme significantly reduces the stronger the integrand oscillates [1]. Thus Filon-type quadratures are highly utile for approximating strongly oscillating integrals with little numerical error.

Although Eq. (5.10) is commonly known, it is simple and instructive to sketch its derivation, so we have it at hand when we turn to the case of higher dimensional quadrature schemes. On each subinterval we wish to interpolate the integral by a three point quadrature scheme

$$\mathbb{I}(x_n, x_{n+2}) = \int_{x_n}^{x_{n+2}} dx\ \varphi(x) e^{i\omega x} \approx w_n \varphi_n + w_{n+1}\varphi_{n+1} + w_{n+2}\varphi_{n+2}. \tag{5.4}$$

Filon's suggestion now results in approximating the preexponential by a second order polynomial $\varphi(x) \approx c_0 + c_1\, x + c_2\, x^2$ which is a good approximation if this function is slowly oscillating as requested. If this approximation of the preexponential was exact, Eq. (5.4) would give an exact value for the integral $\mathbb{I}(x_n, x_{n+2})$. The coefficients w_j can be determined from the integrals over the three moments

$$J_l = \int_{x_n}^{x_{n+2}} dx\ x^l e^{i\omega x} = w_n x_n^l + w_{n+1} x_{n+1}^l + w_{n+2} x_{n+2}^l \qquad l \in \{0, 1, 2\}. \tag{5.5}$$

It is an essential assumption of the Filon method that these moment integrations can be solved analytically. And although there has been much effort to develop moment-free Filon schemes [3], this level of sophistication will suffice for our purposes. Solving the three integrations of Eq. (5.5) for the w_{n+l} we find

$$w_{n+l} = \varsigma_l h\ e^{i\omega x_{n+l}} \tag{5.6a}$$

$$\varsigma_0 = \varsigma_a + i\, \varsigma_b \tag{5.6b}$$

$$\varsigma_a = \frac{3\Theta + \cos(2\Theta)\Theta - 2\sin(2\Theta)}{2\Theta^3} \tag{5.6c}$$

$$\varsigma_b = \frac{2\Theta^2 + 2\cos(2\Theta) + \sin(2\Theta)\Theta - 2}{2\Theta^3} \tag{5.6d}$$

$$\varsigma_1 = \frac{4(\sin(\Theta) - \Theta\cos(\Theta))}{\Theta^3} \tag{5.6e}$$

$$\varsigma_2 = \varsigma_a - i\, \varsigma_b, \tag{5.6f}$$

where one defines the important parameter

$$\Theta = Nh. \tag{5.7}$$

Inserting this into Eq. (5.4) we readily see that this equation now reads

$$\int_{x_n}^{x_{n+2}} dx\varphi(x)\mathrm{e}^{\mathrm{i}\omega x} \approx h\Big[\mathrm{i}\alpha_{\text{Filon}}(\varphi_n \mathrm{e}^{\mathrm{i}\omega x_n} - \varphi_{n+2}\mathrm{e}^{\mathrm{i}\omega x_{n+2}}) \tag{5.8a}$$

$$+ \beta_{\text{Filon}}\left[\varphi_n \mathrm{e}^{\mathrm{i}\omega x_n} + \varphi_{n+2}\mathrm{e}^{\mathrm{i}\omega x_{n+2}}\right] \tag{5.8b}$$

$$+ \gamma_{\text{Filon}}\varphi_{n+1}\mathrm{e}^{\mathrm{i}\omega x_{n+1}}\Big]. \tag{5.8c}$$

where the integration parameters are defined as

$$\alpha_{\text{Filon}} = \frac{2\Theta^2 - 2\sin^2(\Theta) + \sin(2\Theta)\Theta}{2\Theta^3} \tag{5.9a}$$

$$\beta_{\text{Filon}} = \frac{2\Theta(1+\cos^2(\Theta)) - 2\sin(2\Theta)}{2\Theta^3} \tag{5.9b}$$

$$\gamma_{\text{Filon}} = \frac{4(-\Theta\cos(\Theta) + \sin(\Theta))}{\Theta^3}. \tag{5.9c}$$

Summing these contributions over all subintervals we arrive at the expression for the Filon approximation to a univariate integral, which can be found in textbooks as [4]

$$\begin{aligned}\int_a^b dx\, \varphi(x)\,\mathrm{e}^{\mathrm{i}\omega x} &= \sum_{n=0}^{N-1}\int_{x_{2n}}^{x_{2n+2}} dx\, \varphi(x)\,\mathrm{e}^{\mathrm{i}\omega x} \approx \sum_{n=0}^{N-1} \mathbb{A}^{\text{Filon}}(x_{2n}, x_{2n+2}) \\ &= h\Big[\mathrm{i}\alpha_{\text{Filon}}\left(\varphi_0\mathrm{e}^{\mathrm{i}\omega x_0} - \varphi_{2N}\mathrm{e}^{\mathrm{i}\omega x_{2N}}\right) \\ &\quad + \beta_{\text{Filon}}\left(\sum_{n=0}^{N}\varphi_{2n}\mathrm{e}^{\mathrm{i}\omega x_{2n}} - \frac{1}{2}\left[\varphi_{2N}\mathrm{e}^{\mathrm{i}\omega x_{2N}} + \varphi_0\mathrm{e}^{\mathrm{i}\omega x_0}\right]\right) \\ &\quad + \gamma_{\text{Filon}}\sum_{n=1}^{N}\varphi_{2n-1}\mathrm{e}^{\mathrm{i}\omega x_{2n-1}}\Big] \\ &\quad + \mathcal{O}(h^4),\end{aligned} \tag{5.10}$$

with the parameters α_{Filon}, β_{Filon} and γ_{Filon} given by Eqs. (5.9a)–(5.9c).

5.2 Univariate Dynamic Integrals

To obtain an expression as required in Eqs. (5.1) and (5.10) for the univariate dynamic integrals given in Eq. (5.2a), we have to transform that definition into an expression featuring a exponential factor oscillating with a constant frequency. To achieve this goal we employ the fact that the exponential is given by the action Eqs. (3.5), (4.9) and (4.13), whose derivative with respect to the integration variable x^- was shown to never vanish for real x^- (see Sect. 3.3). We can then reparametrize the strictly monotonous exponential phase onto a variable rising linearly to its maximal value

$$S^{k_1}_{p_i,p_f}(x^-) = \upsilon(x^-) S^{k_1,\max}_{p_i,p_f}, \tag{5.11}$$

where $S^{k_1,\max}_{p_i,p_f} = S^{k_1}_{p_i,p_f}(x^-_{\max})$ and $x^-_{\max}$ is defined in Eqs. (5.2a) and (5.2b). By virtue of Eqs. (3.6), (4.10) and (4.14) we find $S^{k_1,\min}_{p_i,p_f} = S^{k_1}_{p_i,p_f}(x^-_{\min}) = 0$ and hence infer that the linear integration variable has to be in the range $\upsilon(x^-) \in [0, 1]$. The univariate dynamic integrals are then to be written in the form

$$f_i = \int_0^1 d\upsilon \frac{dx^-}{d\upsilon} \psi^i_{\mathcal{A}}(x^-(\upsilon)) e^{-i\upsilon S^{k_1,\max}_{p_i,p_f}}, \tag{5.12}$$

The constant oscillation frequency accordingly found in Eq. (5.12) is given by the maximum value of the exponential phase $S^{k_1,\max}_{p_i,p_f}$, and this form is accessible to a numerical quadrature scheme based on Eq. (5.10). The substitution factor is obtained as

$$\frac{dx^-(\upsilon)}{d\upsilon} = \frac{S^{k_1,\max}_{p_i,p_f}}{\partial_{x^-} S^{k_1}_{p_i,p_f}(x^-)}. \tag{5.13}$$

As was argued above it holds $\partial_{x^-} S^{k_1}_{p_i,p_f}(x^-) \neq 0$ for all $x^- \in \mathbb{R}$, whence we see that the factor of Eq. (5.13) is not causing particular difficulties. We finally state that to obtain explicit numerical values of the phase $x^-(\upsilon)$ in dependence of υ, as they enter the shape function $\psi_{\mathcal{A}}(x^-(\upsilon))$ and Eq. (5.13), we are not going to evaluate the implicit relation of Eq. (5.11) in each step, but we will much rather utilize the approximate relation

$$x^-(\upsilon_{n+1}) \approx x^-(\upsilon_n) + \left.\frac{dx^-(\upsilon)}{d\upsilon}\right|_{\upsilon=\upsilon_n} h = x^-(\upsilon_n) + \frac{S^{k_1,\max}_{p_i,p_f}}{\partial_{x^-} S^{k_1}_{p_i,p_f}(x^-(\upsilon_n))} h, \tag{5.14}$$

and the fact that the υ_n are equidistantly spaced, i.e. the step width is given by $h = (x^-_{\max} - x^-_{\min})/2N$. The finally valid formula for approximating the rapidly oscillating integrals of Eq. (3.5), is then given by

$$f_i \approx h\, S^{k_1,\max}_{p_i,p_f} \Bigg[i\alpha_{\text{Filon}} \left(\frac{\psi^i_{\mathcal{A}}(x^-_{\min})}{\partial_{x^-} S^{k_1}_{p_i,p_f}(x^-_{\min})} - \frac{\psi^i_{\mathcal{A}}(x^-_{\max})}{\partial_{x^-} S^{k_1}_{p_i,p_f}(x^-_{\max})} e^{i S^{k_1,\max}_{p_i,p_f}} \right)$$

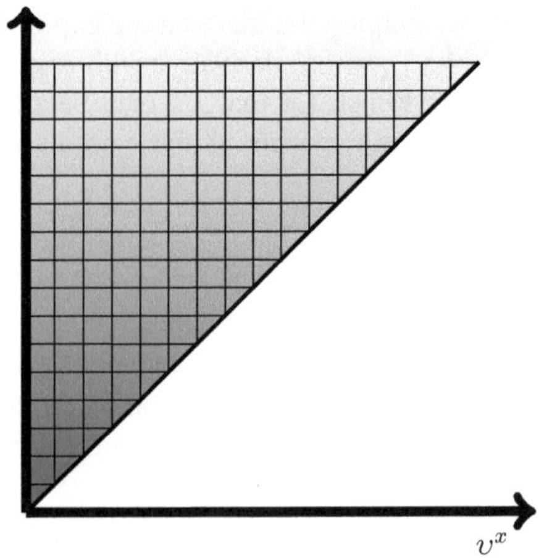

Fig. 5.1 Integration space for bivariate integrals f_{ij} (*shaded area*) and its discretization

$$
\begin{aligned}
&+ \beta_{\text{Filon}} \left(\sum_{n=0}^{N} \left(\frac{\psi^i_{\mathcal{A}}(x^-(\upsilon_{2n}))}{\partial_{x^-} S^{k_1}_{p_i,p_f}(x^-(\upsilon_{2n}))} \mathrm{e}^{\mathrm{i}\upsilon_{2n} S^{k_1,\max}_{p_i,p_f}} \right) \right. \\
&\left. - \frac{1}{2} \left[\frac{\psi^i_{\mathcal{A}}(x^-_{\min})}{\partial_{x^-} S^{k_1}_{p_i,p_f}(x^-_{\min})} + \frac{\psi^i_{\mathcal{A}}(x^-_{\max})}{\partial_{x^-} S^{k_1}_{p_i,p_f}(x^-_{\max})} \mathrm{e}^{\mathrm{i} S^{k_1,\max}_{p_i,p_f}} \right] \right) \\
&+ \gamma_{\text{Filon}} \sum_{n=1}^{N} \frac{\psi^i_{\mathcal{A}}(x^-(\upsilon_{2n-1}))}{\partial_{x^-} S^{k_1}_{p_i,p_f}(x^-(\upsilon_{2n-1}))} \mathrm{e}^{\mathrm{i}\upsilon_{2n-1} S^{k_1,\max}_{p_i,p_f}} \Bigg], \qquad (5.15)
\end{aligned}
$$

with the $2n+1$ equidistantly separated $\upsilon_n = nh$, the step width $h = 1/2N$ and the x^--values obtained from Eq. (5.14). We note that due to the first order approximation, adopted in Eq. (5.14), the estimate for the error to be $\mathcal{O}(h^4)$ is no longer necessarily fulfilled, but it is still negligibly small.

5.3 Bivariate Dynamic Integrals

In principle the univariate integration scheme of the previous section can be generalized to arbitrary highly dimensional integration domains by generalizing Eqs. (5.4) and (5.5) to the requested number of integration variables [5]. This rather tedious and uninstructive procedure, however, will not be followed by us. We will employ special symmetry properties of the bivariate dynamics integrals, whose explicit form is given in Eq. (5.2b). We note that the integrand of the f_{ij} is fully separable into multiplicative functions depending each on only one integration variable x^- or y^-. These integrations can be translated to the scaled variables υ^x, υ^y, each defined in accordance with Eq. (5.13). Due to the step function, relating the two integration dimensions, the integration domain has a triangular shape, as depicted in Fig. 5.1.

The integration over this non-trivially shaped integration domain then calls for a truly multivariate integration scheme. Due to the analogy of each integration dimension to Eq. (5.12) the bivariate dynamic integrals can be written in the form

$$f_{ij} = S^{k_2,\max}_{p_t,p_f} \int_0^1 d\upsilon^y \frac{\psi_{\mathcal{A}}(y^-(\upsilon^y))}{\partial_{y^-} S^{k_2}_{p_t,p_f}(y^-(\upsilon^y))} \mathrm{e}^{-i\upsilon^y S^{k_2,\max}_{p_t,p_f}} \mathbb{I}_x(\upsilon^y) \tag{5.16a}$$

$$\mathbb{I}_x(\upsilon^y) = S^{k_1,\max}_{p_i,p_t} \int_0^{\upsilon^y} d\upsilon^x \frac{\psi_{\mathcal{A}}(x^-(\upsilon^x))}{\partial_{x^-} S^{k_1}_{p_i,p_t}(x^-(\upsilon^x))} \mathrm{e}^{-i\upsilon^x S^{k_1,\max}_{p_i,p_t}} . \tag{5.16b}$$

Here the boundary condition of the integration domain $\upsilon^x < \upsilon^y$ was translated to the condition $\upsilon^{x,\max} = \upsilon^y$. Since in each integration dimension the minimum of the exponential phase is $S^{k_1,\min}_{p_i,p_t} = S^{k_2,\min}_{p_t,p_f} = 0$, we have $\upsilon^{x,\min} = \upsilon^{y,\min} = 0$. The numerical quadrature of Eq. (5.16a) is then analogous to Eq. (5.15), with the complication that the integration domain is not square. This problem, however, is dealt with by splitting the integration domain into many strips in both υ^x- and υ^y-directions. This results in an approximation scheme of the form

$$\begin{aligned} f_{ij} = \sum_{n_y=0}^{N} \sum_{n_x=0}^{n_y} & \int_{\upsilon^y_{2n_y}}^{\upsilon^y_{2(n_y+1)}} d\upsilon^y \frac{\psi_{\mathcal{A}}(y^-(\upsilon^y))}{\partial_{y^-} S^{k_2}_{p_t,p_f}(y^-(\upsilon^y))} \mathrm{e}^{-i\upsilon^y S^{k_2,\max}_{p_t,p_f}} \\ \times & \int_{\upsilon^x_{2n_x}}^{\upsilon^x_{2(n_x+1)}} d\upsilon^x \frac{\psi_{\mathcal{A}}(x^-(\upsilon^x))}{\partial_{x^-} S^{k_1}_{p_i,p_t}(x^-(\upsilon^x))} \mathrm{e}^{-i\upsilon^x S^{k_1,\max}_{p_i,p_t}} \\ + & \Delta(\upsilon^x_{2n_y}, \upsilon^y_{2n_y}), \end{aligned} \tag{5.17}$$

where N is the number of intervals, the integration domain is split up into, in both υ-directions. The symbol $\Delta(\upsilon^x_{2n_y}, \upsilon^y_{2n_y})$ refers to the contribution of the small triangle at the righter edge of every υ^x integration strip (see Fig. 5.1). One of first ideas that one might think of to include these truly two-dimensional portions of the integration domain into the analysis of Eq. (5.17), would be to decrease the step width in each integration dimension until the triangles become small enough that a simple estimate for their contribution, say a complete neglect or approximating them by a square, would result in an acceptable numerical accuracy. This route, however, cannot be successfully taken, since the number of the error-inducing triangles increases linearly with the step-number. The linear decrease of the error in the overall numerical estimate of Eq. (5.17), induced by either one of the above suggested simple approximation schemes for the triangles, is thus compensated by the increased number of separate error sources. In fact, by numerical evaluation it even shows that the error is slightly overcompensated, leading to an increase of the numerical error induced by increasing the number of nodes. This is why the triangles have to be approximated by a true bivariate integration routine. We will employ a simple three-point quadrature scheme in analogy to Eq. (5.4). Since in this case we have to deal with a two dimensional integration domain, we have to modify Eqs. (5.4) and (5.5) for an arbitrary bivariate preexponential function $\varphi(\upsilon^x, \upsilon^y)$ in the following way

$$\int_{\upsilon^y_{2n_y}}^{\upsilon^y_{2(n_y+1)}} \int_{\upsilon^x_{2n_x}}^{\upsilon^y} d\upsilon^y d\upsilon^x \varphi(\upsilon^x, \upsilon^y) e^{i\upsilon^y S^{k_2,\max}_{p_t,p_f}} e^{i\upsilon^x S^{k_1,\max}_{p_i,p_t}}$$

$$\approx w_{2n_y,2n_x} \varphi(\upsilon^y_{2n_y}, \upsilon^x_{2n_x}) e^{i\upsilon^y_{2n_y} S^{k_2,\max}_{p_t,p_f}} e^{i\upsilon^x_{2n_x} S^{k_1,\max}_{p_i,p_t}}$$
$$+ w_{2(n_y+1),2n_x} \varphi(\upsilon^y_{2(n_y+1)}, \upsilon^x_{2n_x}) e^{i\upsilon^y_{2(n_y+1)} S^{k_2,\max}_{p_t,p_f}} e^{i\upsilon^x_{2n_x} S^{k_1,\max}_{p_i,p_t}}$$
$$+ w_{2(n_y+1),2(n_x+1)} \varphi(\upsilon^y_{2(n_y+1)}, \upsilon^x_{2(n_x+1)}) e^{i\upsilon^y_{2(n_y+1)} S^{k_2,\max}_{p_t,p_f}} e^{i\upsilon^x_{2(n_x+1)} S^{k_1,\max}_{p_i,p_t}}, \tag{5.18}$$

where for the oscillation frequencies we already inserted the maxima of the exponential phases and we chose to support the approximation scheme at the three corners of the triangle $\Delta(\upsilon^x_{2n_y}, \upsilon^y_{2n_y})$. To find the prefactors $w_{2(n_y+k),2(n_x+l)}$ we have to solve the three moment equations in the usual way

$$\int_{\upsilon^y_{2n_y}}^{\upsilon^y_{2(n_y+1)}} \int_{\upsilon^x_{2n_x}}^{\upsilon^y} d\upsilon^y d\upsilon^x \left(\upsilon^x\right)^l \left(\upsilon^y\right)^k e^{iS^{k_2,\max}_{p_t,p_f} \upsilon^y} e^{iS^{k_1,\max}_{p_i,p_t} \upsilon^x}$$

$$= w_{2n_y,2n_x} \left(\upsilon^x\right)^l \left(\upsilon^y\right)^k e^{iS^{k_2,\max}_{p_t,p_f} \upsilon^y_{2n_y}} e^{iS^{k_1,\max}_{p_i,p_t} \upsilon^x_{2n_x}}$$
$$+ w_{2(n_y+1),2n_x} \left(\upsilon^x\right)^l \left(\upsilon^y\right)^k e^{iS^{k_2,\max}_{p_t,p_f} \upsilon^y_{2(n_y+1)}} e^{iS^{k_1,\max}_{p_i,p_t} \upsilon^x_{2n_x}}$$
$$+ w_{2(n_y+1),2(n_x+1)} \left(\upsilon^x\right)^l \left(\upsilon^y\right)^k e^{iS^{k_2,\max}_{p_t,p_f} \upsilon^y_{2(n_y+1)}} e^{iS^{k_1,\max}_{p_i,p_t} \upsilon^x_{2(n_x+1)}}, \tag{5.19}$$

with $(l, k) \in \{(0, 0), (0, 1), (1, 0)\}$. Again assuming Eq. (5.19) to be exact and solving for the coefficients $w_{2(n_y+k),2(n_x+l)}$ then gives a Filon-type approximation for a manifestly triangular integration domain. Please note that the special shape of the integration domain is encoded in the boundary conditions for the moment integration and thus implicit in the coefficients $w_{2(n_y+k),2(n_x+l)}$. To obtain an approximation to the remainder of Eq. (5.17), we apply a univariate Filon-type quadrature scheme, as described in the previous section, to the one-dimensional integrals contained in it. For reasons of simplicity we choose to employ the same number of approximation nodes in each dimension, as shown in Fig. 5.1 and the resulting approximation formula is obtained by substituting Eq. (5.15) for the univariate integrals in Eq. (5.17) with the proper replacement of the constants.

References

1. A. Iserles, S.P. Nørsett, Efficient quadrature of highly oscillatory integrals using derivatives. P. Roy. Soc. Lond. A Mat. **461**, 1383 (2005)
2. L.N.G. Filon, On a quadrature formula for trigonometric integrals. Proc. R. Soc. Edinb. **49**, 38 (1928)
3. S. Olver, Moment-free numerical integration of highly oscillatory functions. IMA J. Num. Anal. **26**, 213 (2006)
4. F.W. Olver, D.W. Lozier, R.F. Boisvert, C.W. Clark, in *NIST Handbook of Mathematical Functions*, (Cambridge University Press, New York, 2010)
5. A. Iserles, S.P. Nørsett, Quadrature methods for multivariate highly oscillatory integrals using derivatives. Math. Comp. **75**, 1233 (2006)

Chapter 6
Summary and Outlook

6.1 Summary

The aim of this thesis was to explore to foremost frontiers of strong-field QED in the presence of highly intense laser fields, which exhibit a strong temporal compression. In Chap. 1 it was pointed out, that a natural observable to investigate the underlying physics, are the energy spectra, emitted by electrons when scattered from the considered intense lasers. Thus schemes of describing the effects of temporal focusing of laser pulses on the energy emission spectra of electrons, scattered from such laser pulses, had to be developed.

In Chap. 2 we started with an introduction to the basic concepts of describing the interaction of an electron with a laser wave in classical electrodynamics. Following that we presented a previously undescribed method of exactly modeling the temporal evolution of a spatially focused laser pulse. In particular, this representation can be used to properly account for the spatial focusing of few-cycle laser pulses. In the remainder of Chap. 2 we gave an overview over the basic principles of a quantum electrodynamics in the presence of a strong electromagnetic field, which is not accessible to a perturbative treatment.

Following these introductory remarks, we demonstrated how several fundamental calculational tools of SF-QED, previously studied for the case of a monochromatic external field, can be formally translated to arbitrarily shaped plane wave fields. In the course of these computations we met a variety of computational challenges and possible technical applications of even the two lowest order SF-QED emission processes. We started in Chap. 3 with an investigation of the emission of a single photon by an electron, scattered from an arbitrary plane wave laser field, which we labeled *nonlinear single Compton scattering*. We found that by means of an integration by parts technique, which was demonstrated to be equivalent to gauge invariance, the scattering matrix element turned out to be analytically finite. By virtue of this regularization, the dynamics of the scattering could be subsumed in two one dimensional integrals, called *dynamic integrals*. Evaluating these integrals numerically, we recovered the monochromatic and perturbative limit of nonlinear

K. F. Mackenroth, *Quantum Radiation in Ultra-Intense Laser Pulses*,
Springer Theses, DOI: 10.1007/978-3-319-07740-6_6,

single Compton scattering probability. The analytical details of these limits were moved to Appendix D. Additionally, from a numerical analysis we found that the monochromatic approximation indeed has no predictive power if the external laser field is a few-cycle pulse. In Sect. 3.3 we then thoroughly worked out a stationary phase approximation to the dynamic integrals, valid in the experimentally relevant limit of high electron energies and laser intensities. Based on this stationary phase approximation we established the quasi-classical picture for the electron dynamics in this limit, that it can be quantified in terms of a classical trajectory. On the other hand, we found that the photon emission by the electron is dominated by quantum effects, as soon as the quantum parameter χ reaches order unity. It thus has to be described in a quantum framework. In Sect. 3.4 we employed the analytical results of the stationary phase approximation to propose the first viable scheme of determining the absolute phase of few-cycle laser pulses in the relativistic regime, where conventional determination schemes fail. Having such a scheme available would open access to a deeper understanding of physical processes in such highly intense laser pulses. We finally commented on an old debate about the presence and detectability of a intensity dependent mass dressing of electrons in an intense laser field. We found a corresponding effect, which in addition to the laser intensity depended on several quantities. We also commented on signatures of this mass dressing effect in nonlinear single Compton spectra, whose detection would provide incontestable evidence of mass dressing.

Chapter 4 was devoted to an analysis of the next higher order of the perturbative series of SF-QED. We investigated the scattering matrix amplitude of two photon emission from an electron, scattered from an arbitrarily shaped plane wave field. Analyzing the dressed electron propagator, we found that the scattering matrix element naturally splits up into two parts, which we labeled the *phase ordered* and *same phase* contribution. The physical interpretability of these partial amplitudes was discussed in Sect. 4.4. We noted that only a part of the overall scattering matrix element, connected to the same phase contribution, could be treated in known monochromatic analyses. A contribution, which is connected to the phase ordered partial amplitude and introduces divergences into the monochromatic analysis, was not taken into account. The analytical details of the monochromatic as well as the perturbative limit are investigated in Appendix E. We found the dynamics of the same phase partial amplitude to be governed by one dimensional dynamic integrals, analogous to the results found in Chap. 3, whereas the phase ordered contribution yielded two dimensional dynamic integrals. Analyzing these bivariate dynamic integrals, we demonstrated that the phase ordered contribution is formally equivalent to a phase ordered product of two NSCS amplitudes. Evaluating these dynamic integrals numerically, we found that the phase ordered contribution, not properly described in previous monochromatic analyses, in typical scenarios outscales the same phase contribution by several orders of magnitude and thus largely dominates the overall two photon emission. Apart from this deficiency we found that a monochromatic analysis is not even capable of correctly predicting the mainly emitted photon frequencies, if the external laser field is a few-cycle pulse. In addition to the formal equivalency of the phase ordered partial amplitude to two separate nonlinear single Compton scattering amplitudes we demonstrated that an equivalent result also holds on the probability level. The two single photon emission events, approximately constituing the two photon emission process, were found to be *independent*, if recoil could be neglected. Following this, in Sect. 4.3 we worked out a stationary phase approximation to the scattering matrix element of two photon emission, which analogously to the discussion of Sect. 3.3 allowed for a quasi-classical interpretation of the two photon emission probability, in terms of two smoothly joined classical electron trajectories. This physical picture was then employed in Sect. 4.5 to explain the correlation of the angular emission ranges and the energies of two photons emitted from an electron, scattered from an intense laser field, as the recoil of a photon emission on the electron trajectory and thus every subsequent photon emission.

In Chap. 5 we presented a short overview over the methods employed for obtaining numerical approximations to the dynamic integrals of the previous two chapters. It was demonstrated how the fact, that the dynamic integrals feature no exactly real stationary points, can be exploited to recapture the integrals in terms of a constantly oscillating integrand. This constantly oscillating integrand was then approximated by a standard Filon-type quadrature scheme. For this scheme we presented an explicit two dimensional generalization, applicable to the bivariate dynamic integrals of nonlinear double Compton scattering.

The quantum description of a highly intense, spatially focused laser pulse was investigated in Appendix C. There we presented a solution of the Dirac equation in the presence of a focused Gaussian laser pulse, in terms of an double perturbative series of its exponential phase. We found the intuitive result that in lowest order in the laser focusing the found expression corresponds to the exponential phase of the Volkov solutions of the Dirac equation in a plane wave, with only the plane wave potential replaced by the focused beam potential, depending on all space-time coordinates.

In conclusion we have presented a number of theoretical tools to adapt the available theoretical framework of SF-QED to the challenges of experimental reality.

6.2 Outlook

Looking at the results presented in this thesis, we find several loose ends, which hint at natural continuations of its work. In fact, it is almost as valuable for the prospects it discloses, as its explicit results. We wish to comment on a number of open tasks, the results put forward, this thesis arrived at.

i. Concerning the discussion of Sect. 4.4, that given the explicit representation of the electron propagator in an arbitrarily shaped plane wave potential, there arises the intriguing possibility to generalize the analysis of Chap. 4 to the emission of an arbitrary number of photons. The use of such a scheme to deepen our understanding of fundamental physics, has been highlighted on the example of radiation reaction [1].

ii. The approximation to the wave functions of an electron in the presence of a focused Gaussian beam, found in Appendix C, urge us to translate them into a scattering probability from a spatially focussed laser field. Such an analysis would be of tremendous interest as it would allow for the investigation of the influence of spatial focusing on the expectable radiation patterns in a consistent quantum framework. It is however a highly non-trivial task to solve the resulting multi dimensional dynamic integrals, even when adopting the method of stationary phase.
iii. As is the fate of any theoretical work, anticipating future experimental possibilities also the work presented in this thesis has yet to be subjected to a thorough experimental validation. It is, however, communicated that the corresponding experiments, aiming at the observation of quantum effects predicted in NSCS, are being designed and to be conducted in the near future.

Reference

1. A. Di Piazza, K.Z. Hatsagortsyan, C.H. Keitel, Quantum radiation reaction effects in multiphoton compton scattering. Phys. Rev. Lett. **105**, 220403 (2010)

Appendix A
Differential Probability

The S-Matrix formalism of QED allows us to compute a perturbative expansion of the transition matrix element S_{fi}. To connect this quantity with a physical observable we will always proceed as follows.

It is convenient to observe that the scattering matrix element of a process in a plane wave will always feature a three dimensional energy-momentum conservation. This is cast into the form

$$S_{fi} = \mathcal{N} M_{fi}\, \delta_{\perp}^{(3)} \left(\sum_i p_i - \sum_f p_f \right). \tag{A.1}$$

In this expression the function $\delta_{\perp}^{(3)}$ is the three-dimensional delta function, perpendicular to p_- in the sense explained above.

In any interaction with a plane wave these components are constants of motion whereas p_- is not conserved. The sums inside the perpendicular energy-momentum conservation are over all particles in the initial and final state. In S-Matrix theory now the modulus square of the scattering matrix element S_{fi} gives the probability that the initial state $|i\rangle$ will evolve into the final state $|f\rangle$. Squaring now the δ-function in Eq. (A.1) naively, we obtain

$$\left| \delta_{\perp}^{(3)} \left(\sum_i p_i - \sum_f p_f \right) \right|^2 = \frac{\Delta x^+ L^2}{(2\pi)^3} \delta_{\perp}^{(3)} \left(\sum_i p_i - \sum_f p_f \right), \tag{A.2}$$

where Δx^+ is the dimension of the normalization volume in x^+-direction, whereas L is measured in its canonical directions. Since Δx^+ is not rigorously connected to the canonical normalization volume $V = L^3$, this expression is of little help. To obtain the square of a perpendicular delta function, as found in Eq. (A.1), we thus have to rethink. First we notice that the three-dimensional delta function will always be sufficient to fix all four components of an outgoing particle, since these

K. F. Mackenroth, *Quantum Radiation in Ultra-Intense Laser Pulses*, Springer Theses, DOI: 10.1007/978-3-319-07740-6, have to be physical and their four momentum components thus are connected via the mass-shell condition $p_f^2 = m_f^2$. To avoid the complication of defining the phase space of a particle in light cone coordinates and to obtain a clearly interpretable result, we will consider the state of any final particle in canonical coordinates. Then in the differential transition probability we need the three-dimensional delta function $\delta^{(3)}(\boldsymbol{p}_f - \boldsymbol{p}_f^*)$ with $\boldsymbol{p}_f^*$ being the final particle's canonical spatial momentum. To transform the perpendicular δ-function into an expression in canonical coordinates in a clear and traceable manner, we resort to the following detour, where for reasons of simplicity we assume only one incoming particle. In a realistic scenario we cannot consider the initial particle to be monochromatic in its energy space but much rather have to consider a probability distribution of initial particle momenta and we can write Eq. (A.1) in the form

$$S_{fi} = \mathcal{N} \int d\boldsymbol{p}\, c(\boldsymbol{p})\, M_{fi}(\boldsymbol{p})\, \delta_{\perp}^{(3)}\left(p(\boldsymbol{p}) - \sum_f p_f\right). \tag{A.3}$$

The δ-function can obviously employed to fix the initial particle's momentum and energy to the value

$$\boldsymbol{p}^* = \begin{pmatrix} -\sum_f p_{f,1}^{\perp} \\ -\sum_f p_{f,2}^{\perp} \\ \frac{\left(\sum_f \boldsymbol{p}_f^{\perp}\right)^2 + m^2}{2\sqrt{2}\sum_f \boldsymbol{p}^-} - \frac{\sum_f \boldsymbol{p}^-}{\sqrt{2}} \end{pmatrix} \tag{A.4}$$

$$\varepsilon^* = \frac{\left(\sum_f \boldsymbol{p}_f^{\perp}\right)^2 + m^2}{2\sqrt{2}\sum_f \boldsymbol{p}^-} + \frac{\sum_f \boldsymbol{p}^-}{\sqrt{2}}, \tag{A.5}$$

whereby a conversion factor is introduced into Eq. (A.3) according to

$$\left|\frac{\partial p^-}{\partial p^{\parallel}}\right| = \frac{p^-}{\varepsilon}. \tag{A.6}$$

The square of Eq. (A.3) is then simply

$$\left|S_{fi}\right|^2 = \frac{\mathcal{N}^2}{\left(p^{*-}/\varepsilon^*\right)^2} \left|c(\boldsymbol{p}^*)\right|^2 \left|M_{fi}(\boldsymbol{p}^*)\right|^2. \tag{A.7}$$

To describe the usual situation of a single particle with a fixed energy entering the scattering process we now have to perform the monochromatic limit only on the distribution function of the initial energies

$$c(\boldsymbol{p}) \rightarrow \delta^{(3)}(\boldsymbol{p} - \boldsymbol{p}_i) \tag{A.8}$$

The advantage of this prescription is that in Eq. (A.7) the square $|c(\boldsymbol{p}^*)|^2$ can be readily taken, yielding the well known factor $V/(2\pi)^3$. To obtain a transition probability, however, one has to integrate this result over all possible final states which are available to the final particles. The number of states which are available to the lth final state particle is given by

$$d\Gamma_l = \frac{V \mathrm{d}\boldsymbol{p}_l}{(2\pi)^3}. \tag{A.9}$$

Then to transform the energy momentum conservation law we have found thus far into a condition fixing the spatial momentum of such a final particle, whence we can eliminate the according differential $\mathrm{d}\boldsymbol{p}_l$, we have to find the corresponding energy momentum conservation law arising from the three dimensional δ-function $\delta^{(3)}(\boldsymbol{p}_i - \boldsymbol{p}^*(\boldsymbol{p}_l))$, where we have explicitly written the dependence of $\boldsymbol{p}^*$ and the lth particle's final momentum. The conversion factors are easily found to be

$$\frac{\partial p^{*\perp}_1}{\partial p^{\perp}_{l,1}} = 1 \tag{A.10a}$$

$$\frac{\partial p^{*\perp}_2}{\partial p^{\perp}_{l,2}} = 1 \tag{A.10b}$$

$$\frac{\partial p^{*\parallel}}{\partial p^{\parallel}_l} = \frac{p^-_l}{\varepsilon_l}\frac{\varepsilon_i}{p^-_i} \tag{A.10c}$$

Thus the differential probability, in which the momentum of the jth final particle is fixed, is consequently given by the expression

$$dW = \frac{|\mathcal{N}|^2 V}{(2\pi)^3 \frac{p^-_l}{\varepsilon_l}\frac{p^-_i}{\varepsilon_i}} \left|M_{fi}\right|^2 \delta^{(3)}_{\perp}\left(\boldsymbol{p}_j - \boldsymbol{p}^*_j\right) \Pi_i d\Gamma_i. \tag{A.11}$$

From the normalization of the quantum states given in Chap. 2 we conclude that $\left|M_{fi}\right|^2$ will always contain a factor $(\varepsilon_i \varepsilon_l)^{-1}$, canceling this factor in the denominator of Eq. (A.11). Comparing Eq. (A.11) to the naive guess for the square of the perpendicular δ-function in Eq. (A.2) we are even in the position to state an explicit expression of the before cumbersome ratio

$$\frac{\Delta x^+}{L} = \frac{\varepsilon_i}{p^-_i}. \tag{A.12}$$

We finally state that in actual experiments one does not measure the probability of photon emission given in Eq. (A.11). The experimental data much rather is the energy deposited in the detector by the emitted photons. Angularly and spectrally resolved measurements then yield the differential emitted energy per frequency and solid angle. This quantity, however, is easily obtained from the differential probability from Eq. (A.11), by multiplying it with the energy of the emitted particles. Since the physical scenarios considered in this work refer to the detection of final state photons, the emitted energy is given by

$$dE = \left(\sum_l \omega_l \right) dW(\omega_l), \tag{A.13}$$

where the ω_l are the energies of all emitted photons. In the spectra shown in Chaps. 3 and 4 this quantity is shown, unless explicitly stated otherwise.

Appendix B
The Dirac Equation

In this chapter we wish to compactly summarize the most important features of the Dirac equation, its solutions and the necessary mathematical tools. To this end we largely follow the discussion and conventions of [1]. We recapitulate the wave equation of the Dirac theory, Eq. (2.64a) [2]

$$(\hat{\not{p}} - m)\psi_p = 0. \tag{B.1}$$

In quantum theory the momentum operator is $\hat{p}^\mu = \mathrm{i}\partial^\mu$ and the slash according to Feynman's notation denotes the contraction of a characteristic vector p^μ with a four vector γ^μ of 4×4 matrices, called the *Dirac matrices* and whose anticommutator and commutator are given as

$$\left[\gamma^\mu, \gamma^\nu\right]_+ = 2g^{\mu\nu}, \qquad \sigma^{\mu\nu} = \frac{1}{2}\left(\gamma^\mu\gamma^\nu - \gamma^\nu\gamma^\mu\right). \tag{B.2}$$

From Eq. (B.2) we infer the squares of the Dirac matrices as well as that of a vector contracted with them

$$\left(\gamma^0\right)^2 = -\left(\gamma^1\right)^2 = -\left(\gamma^2\right)^2 = -\left(\gamma^3\right)^2 = \mathbb{1}^4, \qquad \not{a}\,\not{a} = a^2. \tag{B.3}$$

The four vector γ^μ, as any four vector, can be expressed in terms of its time- and space components, as well as in l.c.c.

$$\gamma^\mu = \begin{pmatrix} \gamma^0 \\ \boldsymbol{\gamma} \end{pmatrix} = \begin{pmatrix} \gamma^+ = \frac{\gamma^0 + \gamma^\parallel}{\sqrt{2}} \\ \gamma^- = \frac{\gamma^0 - \gamma^\parallel}{\sqrt{2}} \\ \boldsymbol{\gamma}^\perp \end{pmatrix}, \tag{B.4}$$

where $\gamma^\parallel = \boldsymbol{k}_L\boldsymbol{\gamma}/\omega_L$ and $\boldsymbol{\gamma}^\perp$ are the matrices corresponding to the spatial vector components perpendicular to $\boldsymbol{k}_L$. The conjugate expression of Eq. (B.1) was derived

K. F. Mackenroth, *Quantum Radiation in Ultra-Intense Laser Pulses*,
Springer Theses, DOI: 10.1007/978-3-319-07740-6,

in Sect. 2.2 as the equation of motion of the conjugate field component, resulting from the QED Lagrangian. To derive the same expression from Eq. (B.1) by complex conjugation one concludes that the conjugates of the Dirac matrices are given by

$$\left(\gamma^{\mu}\right)^{\dagger} = \gamma^0\gamma^{\mu}\gamma^0. \tag{B.5}$$

Whenever explicit expressions for the Dirac matrices or their solutions are needed we will adopt the so-called standard representation

$$\gamma^0 = \begin{pmatrix} \mathbb{1}^2 & 0 \\ 0 & -\mathbb{1}^2 \end{pmatrix}, \qquad \boldsymbol{\gamma} = \begin{pmatrix} 0 & \boldsymbol{\sigma} \\ -\boldsymbol{\sigma} & 0 \end{pmatrix}, \tag{B.6}$$

with the three dimensional vector of the 2×2 *Pauli matrices*

$$\sigma^x = \begin{pmatrix} 0 & 1 \\ 1 & 0 \end{pmatrix}, \qquad \sigma^y = \begin{pmatrix} 0 & -\mathrm{i} \\ \mathrm{i} & 0 \end{pmatrix}, \qquad \sigma^z = \begin{pmatrix} 1 & 0 \\ 0 & -1 \end{pmatrix}. \tag{B.7}$$

Equation (B.1) is solved by two types of plane wave solutions

$$\psi_{p,\lambda} = \frac{1}{\sqrt{2\varepsilon V}} u_{p,\lambda} \mathrm{e}^{-\mathrm{i}(px)}, \tag{B.8}$$

$$\psi_{-p,\lambda} = \frac{1}{\sqrt{2\varepsilon V}} u_{-p,\lambda} \mathrm{e}^{\mathrm{i}(px)}, \tag{B.9}$$

where V is a normalization volume and λ is a spin index, discussed further below. The components of the four spinors $u_{p,\lambda}$, $u_{-p,\lambda}$ fulfill the algebraic equation

$$(\not{p} - m)u_{p,\lambda} = 0, \qquad (\not{p} + m)u_{-p,\lambda} = 0 \tag{B.10}$$

and the components of the momentum vector p^{μ} are connected by the requirement $p^2 = m^2$. We adopt the normalization prescriptions

$$\bar{u}_{p,\lambda}u_{p,\lambda} = 2m, \qquad \bar{u}_{-p,\lambda}u_{-p,\lambda} = -2m, \tag{B.11}$$

where $\bar{u}_{\pm p,\lambda} = u^{*}_{\pm p,\lambda}\gamma^0$ means Dirac conjugation. The algebraic relation Eq. (B.10) in standard representation is solved by the spinors

$$u_{\pm p,\lambda} = \begin{pmatrix} \sqrt{\varepsilon \pm m}\, w_{\pm p,\lambda} \\ \sqrt{\varepsilon \mp m}\, (\boldsymbol{n_p}\boldsymbol{\sigma}) w_{\pm p,\lambda} \end{pmatrix}, \tag{B.12a}$$

$$\bar{u}_{\pm p,\lambda} = \left(\sqrt{\varepsilon \pm m}\, w^{*}_{\pm p,\lambda}, -\sqrt{\varepsilon \mp m}\, w^{*}_{\pm p,\lambda}(\boldsymbol{n_p}\boldsymbol{\sigma})\right) \tag{B.12b}$$

with the unit vector $\boldsymbol{n_p} = \boldsymbol{p}/|\boldsymbol{p}|$ and an arbitrary two spinor satisfying the normalization

$$w^{*}_{\pm p,\lambda}w_{\pm p,\lambda} = 1. \tag{B.13}$$

In this quantity the spin states of a free particle are encoded. Since a spin-$1/2$ particle has two possible spin orientations, one has to choose two linearly independent basis vectors $w_{\pm p,\lambda=\pm 1/2}$. For the positive energy solutions a possible choice, corresponding to the projections of the spin on a fixed axis in the particle's rest frame, is given by

$$w_{+p,\lambda=1/2} = \begin{pmatrix} 1 \\ 0 \end{pmatrix}, \qquad w_{+p,\lambda=-1/2} = \begin{pmatrix} 0 \\ 1 \end{pmatrix}. \tag{B.14}$$

We will not need the negative energy two spinors. The above given solutions of the Dirac equation lead to a conserved current of the form

$$\partial_\mu (\bar{\psi}_{\pm p,\lambda} \gamma^\mu \psi_{\pm p,\lambda}) = 0. \tag{B.15}$$

For the constant spinors given in Eq. (B.12b) by multiplying Eq. (B.10) with $\bar{u}_{\pm p,\lambda}$ from the left and extracting a factor p^μ one can easily prove the following important relation

$$\bar{u}_{\pm p,\lambda} \gamma^\mu u_{\pm p,\lambda} = 2p^\mu, \tag{B.16}$$

irrespective of the spin λ. By virtue of the normalization of Eqs. (B.8) and (B.9) this equation leads to a current density of

$$j^\mu_{\pm p} = \bar{\psi}_{\pm p,\lambda} \gamma^\mu \psi_{\pm p,\lambda} = \frac{p^\mu}{\varepsilon V} = \frac{1}{V}(1, \boldsymbol{v}), \tag{B.17}$$

with the particle velocity $\boldsymbol{v} = \boldsymbol{p}/\varepsilon$. We thus infer that the normalization of Eqs. (B.8) and (B.9) corresponds to *one particle per unit volume*. We finally state the important property of the completeness of

$$\sum_\lambda u_{\pm p,\lambda} \bar{u}_{\pm p,\lambda} = \begin{pmatrix} (\varepsilon \pm m)\mathbb{1}^2 & -\boldsymbol{p\sigma} \\ \boldsymbol{p\sigma} & -(\varepsilon \mp m)\mathbb{1}^2 \end{pmatrix} = \not{p} \pm m\mathbb{1}^4, \tag{B.18}$$

which can be checked by direct computation. When referring to the above expressions it is customary to suppress any spin indices as well as unit matrices.

References

1. L.D. Landau and E.M. Lifschitz, Lehrbuch der Theoretischen Physik—Band 4: Quantenelektrodynamik (Akademie Verlag, Berlin, 1991).
2. P.A.M. Dirac, The Quantum Theory of the Electron. P. Roy. Soc. Lond. A Mat. **117**, 610 (1928).

Appendix C
Quantum Scattering Off a Focussed Beam

C.1 Introduction

The key motivation of this work is, to trace the actual trends in high intensity laser physics from a theoretical perspective. We have seen in Chap. 1 that the generation of ever higher laser intensities is experimentally achieved by two major concepts: The temporal and the spatial compression of the available laser energy. In CED the effects of both these trends can be easily traced in the energy emission spectra, by inserting the arbitrarily shaped electromagnetic field into Eq. (2.11) and employ the resulting trajectory in Eq. (2.17). In QED analyses, however, we saw in Sect. 2.2 that one is only able to describe strong laser fields non-perturbatively, if one can provide an exact analytical solution of the Dirac equation Eq. (2.64a). Since such a solution up to date is known only for a plane laser wave (see Eq. (2.98)), spatial focusing cannot be considered in a quantum framework. To overcome this restriction on the available theoretical tool kit, we present a perturbative solution of the Dirac equation in the presence of a focused beam in the following in the frequently encountered parameter regime $\lambda_L \ll w_0$ and $m\xi \ll \varepsilon_i$.

The following discussion will be rather lengthy and technically involved, with the main results summarized in a short paragraph at the end.

C.2 Approximate Solution of the Dirac Equation in a Non-plane Laser Wave

The departing point of the following discussion is the Dirac equation in the presence of a focused external laser field $A_L^\mu(x)$

$$\left[\gamma_\mu \left(\mathrm{i}\partial^\mu - eA_L^\mu(x)\right) - m\right] \Psi_p(x) = 0. \tag{C.1}$$

K. F. Mackenroth, *Quantum Radiation in Ultra-Intense Laser Pulses*,
Springer Theses, DOI: 10.1007/978-3-319-07740-6,

We are going to model the external laser potential by a first order Gaussian beam, according to Eqs. (2.45), (2.58) and (2.60). Since the lowest order of the focusing functions $\Psi^0_{L,n}(\boldsymbol{x})$ are formally equivalent for all field modes, we are going to consider a single mode laser potential $\boldsymbol{A}_L(x) = \boldsymbol{\epsilon}_L \Psi^0_L(\boldsymbol{x})\mathrm{e}^{-\mathrm{i}k_L^+ x^-}$ in the following. The scalar potential can be derived in terms of Eq. (2.46). Following the original derivation of the Volkov solutions of the Dirac equation in a plane wave, we seek for a solution to this equation in the form

$$\Psi_p(x) = \Sigma(x)\,\mathrm{e}^{\mathrm{i}S_p(x)}, \tag{C.2}$$

i.e. a product of a space dependent spinor and an exponential function corresponding to the action of the Volkov solutions. Plugging this ansatz into Eq. (C.1) we arrive at the equation

$$\left[\gamma_\mu\left(\partial^\mu S(x) - eA^\mu_L(x)\right) - m\right]\Sigma_p(x)\,\mathrm{e}^{-\mathrm{i}S_p(x)} = 0, \tag{C.3}$$

where we neglected a term $\gamma_\mu\partial^\mu\Sigma(x)$. The conditions for this approximation to hold, are given below. Using the canonical momentum $\tilde{p}^\mu(x) = \partial^\mu S(x) - eA^\mu_L(x) = (\tilde{\varepsilon}(x), \tilde{\boldsymbol{p}}(x))$ Eq. (C.3) formally turns into a free Dirac equation which is solved by the space dependent spinor

$$\Sigma_p(x) = \begin{pmatrix} \sqrt{\tilde{\varepsilon}(x) + m}\, w \\ \sqrt{\tilde{\varepsilon}(x) - m}\,(\boldsymbol{n}\boldsymbol{\sigma})\, w \end{pmatrix} \tag{C.4}$$

with an arbitrary constant two-spinor w, which can be chosen in accordance to Eq. (B.14), the unit vector in direction of the canonical momentum $\boldsymbol{n} = \tilde{\boldsymbol{p}}/|\tilde{\boldsymbol{p}}|$ and the vector of the Pauli matrices $\boldsymbol{\sigma}$ from Eq. (B.7). Albeit Eq. (C.4) does not constitute an explicit representation of $\Sigma(x)$, with the aid of the canonical momentum $\tilde{p}^\mu(x)$ it can be evaluated at least numerically. To find a solution to Eq. (C.1) it will thus suffice to find a solution $S_p(x)$ of Eq. (C.3). This expression states a system of linear equations which possesses a non-trivial solution if the determinant of the coefficient matrix vanishes. This condition implies

$$\left(\partial^\mu S_p(x) - eA^\mu_L(x)\right)^2 - m^2 = 0. \tag{C.5}$$

Formally this equation is of the type of a classical Hamilton Jacobi equation to find the action $S_p(x)$ of an electron in an electromagnetic potential $A^\mu_L(x)$, whence we will refer to $S_p(x)$ as the action. Analogous to the procedure of Sect. 2.1.3 we are going to look for a solution of Eq. (C.5) in terms of a perturbation series. As a proper expansion parameter for studying a Gaussian beam we have $s = \lambda_L/2\pi w_0 \ll 1$, with the definitions according to Sect. 2.1.3. If $s \ll 1$ is valid, then $A^\mu_L(x)$ is close to a plane wave. It is then advantageous to investigate the scattering in l.c.c. (see Sect. 1.2), since they separate the coordinate a plane wave potential $A^\mu_L(x^-)$ solely depends on, from the three coordinates $x^\perp$ perpendicular to it. From the explicit expression of the four potential of a Gaussian focused beam, given in Eqs. (2.45) and (2.50), we then derive

$$\partial_{x^-} A_L^\mu(x) \sim s\, \nabla^\perp A_L^\mu(x) \sim s^2 \partial_{x^+} A_L^\mu(x) \tag{C.6}$$

with the two dimensional vector of the derivatives with respect to the spatially perpendicular coordinates $\nabla^\perp = (\partial_{x_1^\perp}, \partial_{x_2^\perp})$. Please note that, despite the fact that the approximation of the external laser as a lowest order Gaussian beam never enters the presented computations explicitly, the above scaling laws were derived only for this case. It is presumable that they would unchanged translate to a higher order Gaussian focus, but this has not been investigated yet. As it turns out, inserting for $S_p(x)$ a perturbative series only in the small parameter s leads to differential equations, which are elementary not solvable. We will thus solve Eq. (C.5) for the exponential action in terms of a double sum in orders of s^j as well as the additional small parameter $m\xi/\varepsilon_i \ll 1$. Assuming this ratio to be small, corresponds to the assumption that the electron during the interaction with the laser pulse deviates only slightly from its initial direction, which regime has been thoroughly investigated in SF-QED [1]. The corresponding ansatz for the action of Eq. (C.2) is

$$S_p(x) = -p_\mu x^\mu - \sum_{i,j=0}^{\infty} \left(\frac{m\xi}{\varepsilon}\right)^i s^j\, \chi_{i,j}(x) \tag{C.7}$$

where the action of a free electron $p_\mu x^\mu$ was separated off. Plugging this expansion into Eq. (C.5) one finds a hierarchy of equations, determining the $\chi_{i,j}$. The lowest order equation is the trivial mass-shell condition for the constant momentum p^μ entering Eq. (C.7)

$$p_\mu p^\mu = m^2. \tag{C.8}$$

We may thus interpret p^μ as the action of the electron if the electromagnetic potential goes to zero. We find the higher order terms of Eq. (C.7) by solving the equations

$$\partial_{x^-}\, \chi_{0,0} = k_L^+ \frac{e(pA_L(x))}{pk_L} \tag{C.9a}$$

$$\partial_{x^-}\, \chi_{1,0} = -\frac{k_L^+}{(m\xi)/\varepsilon} \frac{e^2 A_L^\mu(x) A_{L\mu}(x)}{2(pk_L)} \tag{C.9b}$$

$$\partial_{x^-}\, \chi_{0,1} = -\frac{k_L^+}{s(pk_L)} \left(p^\perp \nabla^\perp \chi_{0,0}\right) \tag{C.9c}$$

$$\ldots$$

and so on. These expressions can be integrated to yield the following, albeit lengthy solutions for the terms of the perturbation series $\chi_{i,j}$,

$$\chi_{0,0} = k_L^+ \int_{-\infty}^{x_-} \frac{e(pA_L(c))}{pk_L} dc^- \tag{C.10a}$$

$$\chi_{1,0} = -\frac{k_L^+}{(m\xi)/\varepsilon} \int_{-\infty}^{x^-} \frac{e^2 A_L^\mu(c) A_{L\mu}(c)}{2(pk_L)} dc^- \tag{C.10b}$$

$$\chi_{0,j} = \frac{k_L^+}{s} \int_{-\infty}^{x_-} \frac{\boldsymbol{p}^\perp \partial^\perp \chi_{0,j-1} + 1/s\left(p^+ \partial_{x^+} \chi_{0,j-2}\right)}{pk_L} dc^- \qquad j \geq 1 \tag{C.10c}$$

$$\chi_{i,0} = 0 \qquad i \geq 2 \tag{C.10d}$$

$$\begin{aligned}\chi_{i,j} = &-\frac{k_L^+}{s(pk_L)} \int_{-\infty}^{x^-} \left(\frac{e\boldsymbol{A}_L^\perp(c) \nabla^\perp \chi_{i-1,j-1}}{(m\xi)/\varepsilon} + \boldsymbol{p}^\perp \nabla^\perp \chi_{i,j-1} + \frac{p^+ \partial_{x^+} \chi_{i,j-2}}{s} \right. \\ &\left. + \frac{1}{4(m\xi)/\varepsilon s} \sum_{m,n=0}^{i} \sum_{k,l=0}^{j} \left(\partial_\mu \chi_{m,k}\right)\left(\partial^\mu \chi_{n,l}\right) \delta_{i,m+n+1} \delta_{j,k+l+2} \right) dc^-\end{aligned} \tag{C.10e}$$

Approximating the action by the terms of order s^0 we find

$$S_p(x) \approx -px - k_L^+ \int_{-\infty}^{x^-} e\frac{pA_L(c)}{pk_L} - \frac{e^2 A_L^2(c)}{2(pk_L)} dc^-. \tag{C.11}$$

This is formally equivalent to the phase of the plane wave Volkov solutions (2.98), with the plane wave potential replaced by the focused potential $A_L^\mu(x)$, depending on all space time coordinates. We note that due to Eq. (C.10d) in the limit $s \to 0$ all contributions to $S_p(x)$ proportional to χ_{0j} vanish and the laser potential reduces to the plane wave potential $A_L(x) \to A_L(x^-)$. Then Eq. (C.11) reduces to the Volkov phase from Eq. (2.99).

We will now turn to investigating the contribution from the term neglected in Eq. (C.3). Inserting the found approximate action into the space dependent spinor from Eq. (C.4) we find in the standard representation of the Dirac matrices (see Eq. (B.6))

$$\not{\partial}\Sigma_p(x) = \begin{pmatrix} \left[\frac{\partial_t \tilde{\varepsilon}(x)}{2\sqrt{\tilde{\varepsilon}(x)+m}} - \frac{\nabla\tilde{\varepsilon}(x)\boldsymbol{\sigma}}{2\sqrt{\tilde{\varepsilon}(x)-m}}\right] w \\ \left[\frac{\nabla\tilde{\varepsilon}(x)\boldsymbol{\sigma}}{2\sqrt{\tilde{\varepsilon}(x)+m}} - \frac{\partial_t \tilde{\varepsilon}(x)}{2\sqrt{\tilde{\varepsilon}(x)-m}}\right] w. \end{pmatrix} \tag{C.12}$$

On the other hand we find for the canonical momentum

$$\tilde{\not{p}}(x) - m = \begin{pmatrix} \left[(\tilde{\varepsilon}(x) - m)\sqrt{\tilde{\varepsilon}(x) + m} - \tilde{\boldsymbol{p}}(x)\sqrt{\tilde{\varepsilon}(x) - m}\,\boldsymbol{\sigma}\right] w \\ \left[\tilde{\boldsymbol{p}}(x)\sqrt{\tilde{\varepsilon}(x) + m}\,\boldsymbol{\sigma} - (\tilde{\varepsilon}(x) + m)\sqrt{\tilde{\varepsilon}(x) - m}\right] w. \end{pmatrix} \tag{C.13}$$

In the derivation of Eq. (C.11) we neglected Eq. (C.12) with respect to Eq. (C.13). Since the three Pauli matrices $\boldsymbol{\sigma}$ are linearly independent of the two dimensional unit matrix (see Eq. (B.7)), this neglect is justified if the following inequalities are satisfied

$$\left(\frac{\tilde{\varepsilon}(x)}{m}\right)^2 \gg 1 - \frac{\partial_t \tilde{\varepsilon}(x)}{2m^2} \tag{C.14a}$$

$$\tilde{\boldsymbol{p}}(x) \gg \frac{\nabla \tilde{\varepsilon}(x)}{2(\tilde{\varepsilon}(x) \pm m)}. \tag{C.14b}$$

Since the time derivative of the free electron action $p_\mu x^\mu$ is constant, according to Eq. (C.11) we find

$$\partial_t \tilde{\varepsilon}(x) = \partial_t^2 S_p(t) \approx -\omega_L \partial_t \left(e\frac{pA_L(x^-)}{pk_L} - \frac{e^2 A_L^2(x^-)}{2(pk_L)} \right) \sim \xi. \tag{C.15}$$

The left hand side of Eq. (C.14a), however, is proportional to $\varepsilon/m \gg \xi \gg 1$ and this inequality is satisfied if it holds $\varepsilon \gg m\xi$. A similar argument holds for Eq. (C.14b). We thus find that employing the solution obtained in Eq. (C.11), neglecting the term $\not{\partial}\Sigma_p(x)$ in Eq. (C.3) is always justified in the regime considered in this chapter.

Summarizing the above discussion

We have presented a solution of the Dirac equation in a spatially focused laser field, approximated by a lowest order Gaussian beam. Whereas for the spinors of the solution we only gave an implicit definition, the exponential phase was explicitly derived as a double perturbative series in the two small parameters $s = \lambda_L/2\pi w_0$ (see Sect. 2.1.3) and $m\xi/\varepsilon \ll 1$. The assumption of the latter ratio to be small physically reduces to the assumption that the scattered electron upon interaction with the laser field deviates only little from its initial direction of propagation. We furthermore found the demonstrative result, that in order s^0 the found exponential phase is equivalent to the exponential phase of the Volkov solutions, albeit with the plane wave potential $A_L^\mu(x^-)$ replaced by the focused potential $A_L^\mu(x)$, depending on all space time coordinates. This leads to the required property of the found phase, that in the limit $s \to 0$, corresponding to the transition of the scattering laser field to a plane wave, it reproduces the well known Volkov result.

In principle it is now possible to obtain scattering amplitudes for arbitrary SF-QED processes in the presence of a focused Gaussian beam. A strong complication of this task, however, is the fact that in these calculations no coordinate dependency can be integrated out trivially. The dynamic integrals of the process will then be over truly four dimensional, highly oscillating integrands. The numerical quadrature of such expressions is highly non-trivial and well beyond the scope of this thesis. An approximate evaluation of SF-QED scattering amplitudes in the presence of a strong focused Gaussian beam in terms of the stationary phase method presented in Sects. 3.3 and 4.3 seems to be closer in reach. Evaluating the stationary point condition Eq. (3.24) beyond lowest order approximation of the exponential phase, found in this chapter, however, requires an analytical solution of a four dimensional differential equation. This computation remains an interesting, but still open task.

Reference

1. V.N. Baier, V.M. Katkov, V.M. Strakhovenko, *Electromagnetic Processes at High Energies in Oriented Single Crystals* (World Scientific, Singapore, 1994)

Appendix D
Asymptotic Limits of Nonlinear Single Compton Scattering

In this chapter we will work out the analytic details of asymptotic limits of NSCS, mentioned in Chap. 3.

D.1 Perturbative Limit

The notion *perturbative limit* refers to the parameter regime, in which the results of the ordinary QED perturbation theory are recovered. To allow for this recovery, the expansion parameter of the perturbation series in interactions with the external laser wave has to be small. As was mentioned in Chap. 1 this implies the limit $\xi \ll 1$. Then the scattering laser field is weak and one can view the interaction with it as a single photon exchange. Neglecting terms of order ξ^2 or higher in Eq. (3.3), S_{fi} must consequently coincide with the lowest order scattering matrix element for linear Compton scattering, provided we replace the normalization of the laser's amplitude with the appropriate single photon normalization factor. From Eq. (3.3) we obtain the $\mathcal{O}(\xi)$ contributions to the scattering matrix element as

$$S_{fi} \approx -\mathrm{i}\frac{e\sqrt{4\pi}(2\pi)^3}{\sqrt{8\omega_1\,\varepsilon_f\,\varepsilon_i\,V^3}} \int \mathrm{d}x^- \bar{u}_{p_f}\Gamma_1 u_{p_i}\psi_{\mathcal{A}}(x^-)\mathrm{e}^{-\mathrm{i}\gamma_V k_L^+ x^-}\delta^{(3)}\left(p_i^\perp - k_1^\perp - p_f^\perp\right). \tag{D.1}$$

To evaluate the contained integral, we expand the shape function as a Fourier integral

$$\psi_{\mathcal{A}}(x^-) = \int_{-\infty}^{\infty} \mathrm{d}\varpi c(\varpi)\mathrm{e}^{-\mathrm{i}\,\varpi\,k_L^+ x^-} \tag{D.2a}$$

$$\text{with } c(\varpi) = \frac{k_L^+}{2\pi}\int_{-\infty}^{\infty} \mathrm{d}c^-\;\psi_{\mathcal{A}}(c^-)\mathrm{e}^{\mathrm{i}\,\varpi\,k_L^+ c^-}. \tag{D.2b}$$

K. F. Mackenroth, *Quantum Radiation in Ultra-Intense Laser Pulses*,
Springer Theses, DOI: 10.1007/978-3-319-07740-6,

The reason for scaling the transformed variable ϖ by k_L^+ is that defined in this fashion it becomes dimensionless and independent of the laser's central frequency ω_L and can consequently be understood as a frequency ω, contained in the pulse's Fourier decomposition, in units of ω_L. The laser pulse is then to be understood as a continuous composition of field modes with wave vectors $k_L^\varpi = \varpi k_L^\mu$. From these definitions we obtain by partial integration

$$\frac{f_1}{\gamma_V} = -\int_{-\infty}^{\infty} \mathrm{d}x^- \mathrm{d}\varpi \, \frac{c(\varpi)}{\varpi} \mathrm{e}^{-\mathrm{i}(\gamma_V+\varpi)k_L^+ x^-} . \tag{D.3}$$

The function $c(\varpi)$ essentially is the Fourier transformation of pulse's electric field, whence we conclude that the factor $|c(\varpi)|^2$ is a weighting factor, giving the relative strength of frequency component $\varpi\omega_L$ to the laser field. Plugging Eq. (D.3) into Eq. (D.1) and using Eqs. (D.2a), after some algebraic transformations we find for the scattering matrix element

$$S_{fi} \approx \int_0^\infty \mathrm{d}\varpi \, c(\varpi) \mathcal{N}_{\text{pert}}^{\text{NSCS}}(\varpi) \bar{u}_{p_f} \left[M_{fi}^{(1)}(\varpi) + M_{fi}^{(2)}(\varpi) \right] u_{p_i} \delta^{(4)} \left(p_i + k_L^\varpi - k_1 - p_f \right) . \tag{D.4}$$

Since for $\varpi < 0$ the energy momentum conservation of Eq. (E.9) cannot be fulfilled, we omitted all those Fourier components. The normalization constant and the reduced matrix elements are given by

$$\mathcal{N}_{\text{NSCS}}^{\text{pert}}(\varpi) = -\,\mathrm{i} \frac{e^2 A_L^\varpi \sqrt{4\pi} (2\pi)^4}{\sqrt{8\omega_1 \, \varepsilon_f \, \varepsilon_i \, V^3}} \tag{D.5a}$$

$$M_{fi}^{(2)}(\varpi) = -\frac{\not{\epsilon}_L \left(\not{p}_f - \not{k}_L^\varpi + m \right) \not{\epsilon}_1^*}{2(p_f k_L^\varpi)}, \qquad M_{fi}^{(1)}(\varpi) = \frac{\not{\epsilon}_1^* \left(\not{p}_i + \not{k}_L^\varpi + m \right) \not{\epsilon}_L}{2(p_i k_L^\varpi)}. \tag{D.5b}$$

We made use of the algebraic relation Eq. (B.10) and A_L^ϖ is the now frequency dependent normalization factor of the photon field. For this normalization we now have to insert the normalization constant for a single photon of frequency $\varpi\omega_L$, which is given by $A_L = \sqrt{4\pi/2\varpi\omega_L V}$, instead of the laser wave normalization $A_L = m\xi/|e|$. The two terms of Eq. (D.5a) correspond to the two possible channels of perturbative Compton scattering depicted in Fig. (D.1). The integrand in Eq. (D.4) is then exactly equivalent to the perturbative result for single photon Compton scattering [1] integrated over the frequency components of the scattering field each weighted with the Fourier coefficient $c(\varpi)$. This result is a generalization of the textbook result for a broad spectrum of incident photon frequencies, which to the best of the author's knowledge was not reported before. It was, however, expected, since the amplitude for the linear Compton effect is linear in the interaction term $e(jA_L)$. Hence for a

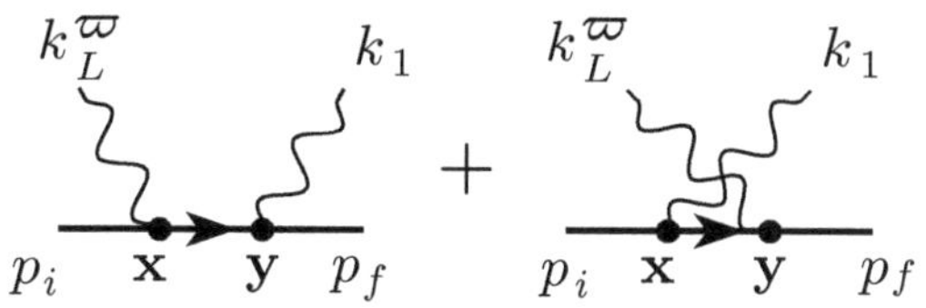

Fig. D.1 Two possible channels of perturbative Compton scattering

non-singular distribution of frequencies contained in A_L^μ it has to be proportional to the Fourier integral over all contained frequency components.

From Eq. (D.4) one can easily obtain the text-book result of Compton scattering from a photon of fixed energy. To consider such a singular photon frequency distribution, we have to compute the Fourier distribution $c(\varpi)$ corresponding to a single photon wave function for the incoming photon. From Eq. (D.2b) we conclude this to be given by

$$c^{\text{m.c.}}(\varpi) = \frac{k_L^+}{2\pi} \int_{-\infty}^{\infty} \mathrm{d}\tilde{x}\, \mathrm{e}^{-\mathrm{i}k_L^+\tilde{x}} \mathrm{e}^{\mathrm{i}\,\varpi k_L^+\tilde{x}^-} = -\mathrm{i}\delta\left(\varpi - 1\right). \tag{D.6}$$

Inserting this expression into Eq. (D.4) we recover the exact single photon Compton scattering matrix element [1]. To obtain its modulus square and hence the corresponding scattering cross section the usual spin and polarization summing and averaging techniques are applied. We only give the final result given, as is customary, in the reference frame in which the electron initially is at rest ($p_i = m(1, 0, 0, 0)$) [1, 2]

$$\frac{\mathrm{d}\sigma}{\mathrm{d}\Omega_1} = \frac{r_e^2}{2} \left(\frac{\omega_1}{\omega_L}\right)^2 \left[\frac{\omega_L}{\omega_1} + \frac{\omega_1}{\omega_L} - \sin^2\theta\right], \tag{D.7}$$

with θ the angle between the vectors $\boldsymbol{k}_1$ and $\boldsymbol{k}_L$. The result of Eq. (D.7) is the well known Klein-Nishina formula for single photon Compton scattering. The same result of course is obtained if one takes the perturbative limit as a second step from the monochromatic limit of the complete emission probability. This latter limit shall be investigated in the following section.

D.2 Monochromatic Limit

Since it has been the main focus of strong-field QED for several decades, we wish to shortly sketch the limiting case of a monochromatic external laser wave [3]. Furthermore this task is not trivial, as it requires care in taking the square of involved δ-functions which is not discussed in the literature, so far (compare Appendix A). As was shown in Sect. 2.2, in a monochromatic plane wave laser field the Volkov wave functions can be expanded in a Fourier-type series of $n \in [0, \infty]$ absorbed or emitted photons. The resulting structure of the scattering matrix element, which involves all space time coordinates only as exponential phases, will be proportional to a four-dimensional energy-momentum conserving δ-function. This behaviour, however, can be directly obtained from the result Eq. (3.3), inserting a monochromatic shape function. In accordance with Sect. 2.2 we consider the sine $\psi_{\mathcal{A}}(\eta) = \sin(\eta)$ (the analysis for $\psi_{\mathcal{A}}(\eta) = \cos(\eta)$ is analogous). Inserting this shape function into Eq. (3.5), and again employing the generating functional relation of the special Bessel functions Eq. (2.116) we obtain the expression

$$f_i = \sum_{n=-\infty}^{\infty} \int \mathrm{d}x^- \; C_{i,n}\, \mathrm{e}^{-i(q_i^+ + nk_L^+ - k_1^+ - q_f^+)x^-} = 2\pi \sum_{n=-\infty}^{\infty} C_{i,n}\, \delta(q_i^+ + nk_L^+ - k_1^+ - q_f^+). \tag{D.8}$$

In this expressions the dressed momenta $q^\mu = p^\mu + \left(m^2\xi^2/2\right) k_L^\mu$ had to be substituted for the free momenta, as shown in Sect. 2.2. To compute the monochromatic emission probability from Eq. (D.8), we have to take the square of this form of the dynamic integrals. In principle, following the arguments of Appendix A, the square of a δ-function is formally taken by considering its generating integral in a finite volume and in the end letting the volume tend to infinity. Applying this procedure in Eq. (D.8) results in the expressions

$$|f_i|^2 = (2\pi)\Delta x^- \sum_{n=-\infty}^{\infty} \left|C_{i,n}\right|^2 \delta(q_i^+ + nk_L^+ - k_1^+ - q_f^+) \tag{D.9a}$$

$$\mathrm{Re}\left(f_i f_j^*\right) = (2\pi)\Delta x^- \sum_{n=-\infty}^{\infty} \mathrm{Re}(C_{i,n}C_{j,n}{}^*)\delta(q_i^+ + nk_L^+ - k_1^+ - q_f^+). \tag{D.9b}$$

Please note that the actual double sum in this expression is naturally collapsed by the δ-functions, which share the same argument. In the ab-initio monochromatic analysis, this additional energy-momentum conservation law is translated into a law confining the frequency of the emitted photon, where the actually diverging total emission probability is taken per unit time, i.e., divided by the time interval T, and hence finite. The emission probability derived from Eqs. (D.9a) will then contain a factor $\Delta x^-/T$. This additional factor is easily determined by resorting to the condition of a conserved four-volume expressed in Eq. (1.18). The accordingly needed coordinate quotient can be obtained from the initial analysis of Appendix A, which in the monochromatic case has to be carried out in terms of dressed momenta and energies. We finally arrive at the result

$$\frac{\Delta x^-}{T} = \frac{q_i^-}{q_i^0}, \tag{D.10}$$

where $q_i^- = p_i^-$ (see Eq. (2.114a)). We use Eqs. (D.9a) and (D.10) plugged into Eq. (3.21) to find the emission probability

$$\frac{\mathrm{d}W^{\text{m.c.}}}{T\mathrm{d}\omega_1\mathrm{d}\Omega_1}=\frac{e^2\,\omega_1}{16\pi q_i^0 p_f^-}\sum_{n=-\infty}^{\infty}\sum_{\{\sigma,\lambda\}}\left|\Gamma_1 A_n^1+\Gamma_2 A_n^2\right|^2\delta(q_i^+ + nk_L^+ - k_1^+ - q_f^+). \tag{D.11}$$

As is customary in monochromatic analyses the additional δ-function is employed to fix the emitted photon's frequency ω_1. To this end we need to find the derivative of its argument with respect to ω_1

$$\left|\frac{\mathrm{d}(q_i^+ + nk_L^+ - k_1^+ - q_f^+)}{\mathrm{d}\omega_1}\right|=\frac{(q_i+nk_L)n_1}{p_f^-}=\frac{q_f^0}{p_f^-}\left|\frac{\mathrm{d}(q_f^0+\omega_1)}{\mathrm{d}\omega_1}\right|. \tag{D.12}$$

We consequently obtain as emission probability per unit solid angle and time

$$\frac{1}{T}\frac{\mathrm{d}W^{\text{m.c.}}}{\mathrm{d}\Omega_1}=\frac{e^2\,\omega_1}{16\pi q_i^0 q_f^0\left|\frac{\mathrm{d}(q_f^0+\omega_1)}{\mathrm{d}\omega_1}\right|}\sum_{n=-\infty}^{\infty}\sum_{\{\sigma,\lambda\}}\left|\Gamma_1 C_{1,n}+\Gamma_2 C_{2,n}\right|^2. \tag{D.13}$$

Recalling that the renormalization scheme of Eq. (3.13) also holds in the monochromatic limit, we find this expression to agree with Eq. (2.127). The cross section can be obtained dividing Eq. (D.13) by the relative flux of incident particles. The subtleties of this procedure, however, are widely discussed in the literature [3–6] and we do not elaborate on them at this point.

D.3 Classical Limit

In Chap. 1 it was stated, that the only quantum effect, affecting the energy emission spectra, is the photon recoil. If one considers optical laser frequencies, one thus expects an agreement between the quantum analysis and the classical results in the limit

$$\frac{(k_L k_1)}{(k_L p_i)}\ll 1. \tag{D.14}$$

To implement this limit in the obtained quantum result, we will not depart from Eq. (3.3), where the matrix factors of the Volkov wave functions have already been expanded. We rather start from the original expression Eq. (3.2). Integrating out the perpendicular momentum components as usual we find

$$S_{fi}=-\mathrm{i}\frac{\sqrt{(4\pi)}e(2\pi)^3}{\sqrt{2\omega_1 V^3}}\int dx^-\,\overline{\Psi}_{p_f}(x^-)\not{\epsilon}_1^*\mathrm{e}^{\mathrm{i}k_1 x}\Psi_{p_i}(x^-)\delta^{(3)}(p_i^\perp - p_f^\perp - k_1^\perp). \tag{D.15}$$

The Ritus matrices entering $\Psi_{p_i,p_f}(x^-)$ depend on the momenta only via the combination $p_{i,f}k_L$, whence we conclude in the classical limit it must hold $E_{p_f}\approx E_{p_i}$. In

Sect. 2.2, however, it was shown $\overline{E}_{p_i,\sigma}\gamma^\mu E_{p_i,\lambda} = j^\mu_{\sigma,\lambda}$, where σ, λ explicitly denote spin states. The exponential phase, which is given by Eq. (3.12), in the classical limit is found to depend on the limits of the Volkov parameters

$$\tilde{\alpha}_V \approx e\frac{k_1 A_L}{k_L p_i} \tag{D.16a}$$

$$\tilde{\beta}_V \approx \frac{e^2 A_L^2 (k_L k_1)}{2(k_L p_i)^2} \tag{D.16b}$$

$$\tilde{\gamma}_V = -\frac{k_1 p_i}{k_L p_i}. \tag{D.16c}$$

The last exact equality is not affected by the limit Eq. (D.14). Comparing these factors to Eq. (2.34) we find that the action in the classical limit is equivalent to the exponent of the classical radiation formula Eq. (2.17)

$$k_1 x(x^-) = -k_L^+ \int_{-\infty}^{x^-} dc^- \tilde{\alpha}_V \psi_A(c^-) + \tilde{\beta}_V \psi_A^2(c^-) + \tilde{\gamma}_V \tag{D.17}$$

A simplified form of the classical limit of Eq. (3.3) is thus found as

$$S_{fi} = -\mathrm{i}\frac{(4\pi)e^2(2\pi)^3}{\sqrt{2\omega_1 V^3}}\,\epsilon^*_{1\,\mu} \int dx^-\, j^\mu_{\sigma,\lambda}(x^-)\delta_{\sigma,\lambda} \mathrm{e}^{\mathrm{i}S(x^-)}\delta^{(3)}(p_i^\perp - p_f^\perp - k_1^\perp), \tag{D.18}$$

where according to Eq. (D.17) the action $S(x^-)$ is equivalent to $k_1 x(x^-)$. Here the Kronecker symbol $\delta_{\sigma\lambda}$ denotes that the quantum current $j^\mu_{\sigma,\lambda}(x^-)$ vanishes for $\sigma \neq \lambda$, as is seen from the explicit representations in Eq. (B.14). This fact simply reflects that an electron, propagating in a plane wave laser field, does not change its spin. From this observation we conclude that the spin sum and average over the outgoing and incoming electron spin states, respectively, in Eq. (D.18) reduces to $1/2\sum_{\sigma,\lambda} j^\mu_{\sigma,\lambda}(x^-)\delta_{\sigma,\lambda} = j^\mu_{\sigma,\sigma}(x^-) =: j^\mu(x^-)$. The sum over the outgoing photon's polarization states is equally easy and for the modulus square of Eq. (D.18) one obtains

$$\left|S_{fi}\right|^2 = \frac{e^2(2\pi)^7}{\omega_1 V^3}\left|\int dx^-\, j^\mu(x^-)\mathrm{e}^{\mathrm{i}S(x^-)}\delta^{(3)}(p_i^\perp - p_f^\perp - k_1^\perp)\right|^2. \tag{D.19}$$

This expression is already reminiscent of Eq. (2.17), however, one has to be careful since the classical current $j^\mu(x^-) = p^\mu(x^-)/\varepsilon(x^-)$ is not equal to the quantum current $j^\mu(x^-) = \overline{\Psi}_{p_i}\gamma^\mu\Psi_{p_i} = p^\mu(x^-)/\varepsilon_i$. The square of the perpendicular δ-functions in Eq. (D.19) is outlined in Appendix A. Then by inserting the expression for the quantum current (Eq. (2.105)) we find for the classical limit of the energy spectrum

$$\begin{aligned}\frac{dE}{d\omega_1 \mathrm{d}\Omega_1} &= |S_{fi}|^2 \frac{\omega_1^3 V^2}{(2\pi)^6} \mathrm{d}^3\boldsymbol{p}_f \\ &= \frac{e^2\omega_1^2}{(2\pi)^2 p_i^- p_f^-} \left| \int dx^- \, p^\mu(x^-) \mathrm{e}^{\mathrm{i}S(x^-)} \right|^2,\end{aligned} \tag{D.20}$$

where $p^\mu(x^-)$ is the classical phase dependent electron momentum. Given that due to Eq. (D.14) in the classical limit it holds $p_i^- \approx p_f^-$, this expression is equivalent to Eq. (2.35).

References

1. L.D. Landau and E.M. Lifschitz, Lehrbuch der Theoretischen Physik—Band 4: Quantenelektrodynamik (Akademie Verlag, Berlin, 1991).
2. F. Mandl, G. Shaw, *Quantum field theory* (John Wiley & Sons, New York, 1984)
3. V.I. Ritus, Quantum effects of the interaction of elementary particles with an intense electromagnetic field. J. Russ. Laser Res. **6**, 497 (1985)
4. L.S. Brown, T.W.B. Kibble, Interaction of Intense Laser Beams with Electrons. Phys. Rev. **133**, A705 (1964)
5. A.I. Nikishov, V.I. Ritus, Quantum processes in the field of a plane electromagnetic wave and in a constant field.1. Sov. Phys. JETP **19**, 529 (1964)
6. Z. Fried, A. Baker, D. Korff, Comments on intensity-dependent frequency shift in Compton scattering and its possible detection. Phys. Rev. **151**, 1040 (1966)

Appendix E
Asymptotic Limits of Nonlinear Double Compton Scattering

In this chapter we will work out some analytic details of NDCS, mentioned in Chap. 4. We begin with a derivation of the step function, employed in deriving Eq. (4.37), and afterwards turn to the asymptotic limits of Eq. (4.25) in the perturbative and monochromatic case.

E.1 Integral Representation of the Step Function

The following is to be understood in terms of distributions. However, according to physicists' conventions we will write functions, keeping in mind the mathematical complications of this term. The Heaviside step function is representable as an integral over a delta function

$$\Theta(x) = \int_{-\infty}^{x} \delta(x')dx'. \tag{E.1}$$

Inserting the Fourier transform of $\delta(x')$ it is found

$$\begin{aligned}\Theta(x) &= \frac{1}{2\pi}\int_{-\infty}^{\infty}\int_{-\infty}^{x} \mathrm{e}^{\mathrm{i}px'}\,dx'dp \\ &= \frac{1}{2\pi\mathrm{i}}\int_{-\infty}^{\infty} \lim_{\epsilon\to 0}\left.\frac{\mathrm{e}^{\mathrm{i}(p-\mathrm{i}\epsilon)x'}}{p-\mathrm{i}\epsilon}\right|_{x'=-\infty}^{x'=x} dp \\ &= \frac{1}{2\pi\mathrm{i}}\int_{-\infty}^{\infty} \mathcal{P}\left(\frac{\mathrm{e}^{\mathrm{i}px}}{p}\right) dp.\end{aligned} \tag{E.2}$$

K. F. Mackenroth, *Quantum Radiation in Ultra-Intense Laser Pulses*, Springer Theses, DOI: 10.1007/978-3-319-07740-6,

Here the principal value prescription is necessarily introduced since the point $p = 0$ needs to be explicitly excluded, for the presented transformations to be meaningful. To invert this expression, aiming at a representation of the right hand side integral in terms of the step function, we need to also consider the case $x < 0$ which is not excluded in the integral expression. By means of a sign change of the integration variable in Eq. (E.2) it is shown

$$\Theta(-x) = -\frac{1}{2\pi \mathrm{i}} \int_{-\infty}^{\infty} \mathcal{P}\left(\frac{\mathrm{e}^{\mathrm{i}px}}{p}\right) dp. \tag{E.3}$$

Combining Eqs. (E.2) and (E.3) thus gives

$$\int_{-\infty}^{\infty} \mathcal{P}\left(\frac{\mathrm{e}^{\mathrm{i}px}}{p}\right) dp = \mathrm{i}\pi \operatorname{sgn}(x), \tag{E.4}$$

with the sign function $\operatorname{sgn}(x) = \Theta(x) - \Theta(-x)$.

E.2 Perturbative Limit

In analogy to the discussion of Sect. D.1 we expect that in the limit $\xi \ll 1$ the electron interacts with only single photons from the external laser wave and thus the results of the usual QED perturbation theory are restored. From Eq. (4.17) we infer that the phase ordered partial amplitude is of order $\sim \xi^2$ in lowest order and will thus not contribute to the perturbative limit. This observation is consistent with the interpretation that the phase ordered partial process corresponds to two separate single photon emissions which each is only possible if at least one photon from the laser field is absorbed. The $\mathcal{O}(\xi)$ expansion of $S_{fi}^{\text{s.p.}}$ is given by

$$S_{fi} \approx \mathcal{N}_{\text{NDCS}} \int_{-\infty}^{\infty} dx^- \bar{u}_{p_f} \frac{\Gamma_1}{2(p_t k_L)} u_{p_i} \psi_{\mathcal{A}}(x^-) \mathrm{e}^{-\mathrm{i}\gamma_V x^-} \delta^{(3)}(p_i^\perp - k_1^\perp - k_2^\perp - p_f^\perp) \tag{E.5}$$

with the definition of $\mathcal{N}_{\text{NDCS}}$, Γ_1 and γ_V according to Eqs. (4.11), (4.23) and (4.24), respectively. In the following we resort to a continuous Fourier decomposition of the shape function in analogy to Sect. D.1

$$\psi_{\mathcal{A}}(x^-) = \int_{-\infty}^{\infty} d\varpi c(\varpi) \mathrm{e}^{-\mathrm{i}\varpi k_L^+ x^-} \tag{E.6}$$

$$c(\varpi) = \frac{k_L^+}{2\pi} \int_{-\infty}^{\infty} dc^- \ \psi_{\mathcal{A}}(c^-) e^{i \varpi k_L^+ c^-}. \tag{E.7}$$

To remove the inverse of the Volkov factors $\gamma_V^{x,y}$, appearing in the definition of Γ_1, we employ a partial integration, analogous to Eq. (D.3), resulting, after some algebra and use of Eq. (E.6), in the replacements

$$\frac{\not{k}_L + \frac{\not{p}_t + m}{\gamma_V^x}}{2(k_L p_t)} f_1 = -(2\pi) \int_{-\infty}^{\infty} d\varpi c(\varpi) \frac{\not{p}_i - \not{k}_1 + m}{2(p_i k_1)} \delta(p_i^+ + k_L^{\varpi +} - k_1^+ - p_f^+ - k_2^+) \tag{E.8a}$$

$$\frac{\not{k}_L - \frac{\not{p}_t + m}{\gamma_V^y}}{2(k_L p_t)} f_1 = (2\pi) \int_{-\infty}^{\infty} d\varpi c(\varpi) \frac{\not{p}_f + \not{k}_2 + m}{2(p_f k_2)} \delta(p_i^+ + k_L^{\varpi +} - k_1^+ - p_f^+ - k_2^+). \tag{E.8b}$$

These relations are inserted into Eq. (E.5). The scattering matrix element then after some algebraic transformations, which makes use of Eq. (B.10) and the three dimensional energy momentum conservation, is expressible as

$$S_{fi} = \mathcal{N}_{\text{NDCS}}^{\text{pert}} \int_0^{\infty} d\varpi c(\varpi) \bar{u}_f \sum_{i=1}^{3} M_{fi}^i(\varpi) u_i \ \delta^{(4)}(p_i + k_L^{\varpi} - k_1 - k_2 - p_f) \tag{E.9}$$

$$\mathcal{N}_{\text{NDCS}}^{\text{pert}} = -i \frac{(2\pi)^5 e^3 A_L}{\sqrt{4\omega_1 \omega_2 \varepsilon_i \varepsilon_f V^4}},$$

where as discussed in Sect. D.1 all frequency components $\varpi < 0$ are omitted, $k_L^{\varpi} = \varpi k_L^{\mu}$ and the reduced matrix elements read

$$M_{fi}^1(\varpi) = \not{\epsilon}_L \frac{\not{p}_f - \not{k}_L^{\varpi} + m}{2(k_L^{\varpi} p_f)} \not{\epsilon}_2^* \frac{\not{p}_i - \not{k}_1 + m}{2(k_1 p_i)} \not{\epsilon}_1^* \tag{E.10a}$$

$$M_{fi}^2(\varpi) = \not{\epsilon}_2^* \frac{\not{p}_f + \not{k}_2 + m}{2(k_2 p_f)} \not{\epsilon}_1^* \frac{\not{p}_i + \not{k}_L^{\varpi} + m}{2(k_L^{\varpi} p_f)} \not{\epsilon}_L \tag{E.10b}$$

$$M_{fi}^3(\varpi) = -\not{\epsilon}_2^* \frac{\not{p}_f + \not{k}_2 + m}{2(k_2 p_f)} \not{\epsilon}_L \frac{\not{p}_i - \not{k}_1 + m}{2(k_1 p_i)} \not{\epsilon}_1^*. \tag{E.10c}$$

These reduced matrix elements are equivalent to the analytical expressions resulting from the corresponding Feynman diagrams for perturbative double Compton scattering, obtained from ordinary QED, (see Fig. (E.1)) [1].

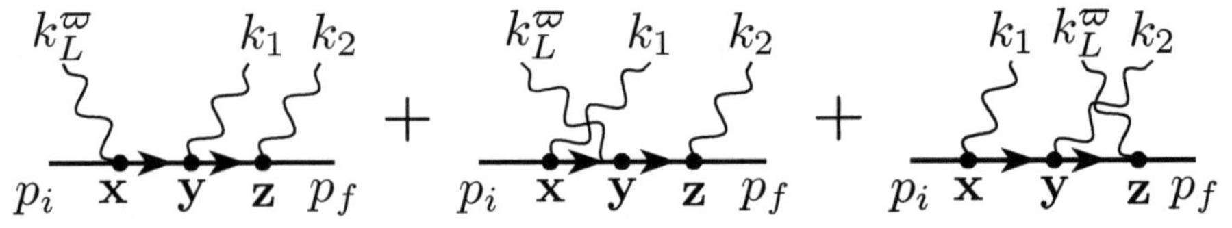

Fig. E.1 Three possible channels of single photon double Compton scattering

E.3 Monochromatic Limit

To obtain the monochromatic limit, we consider shape function

$$\psi_{\mathcal{A}}(x^-) = \sin(k_L^+ x^-). \tag{E.11}$$

To investigate the effect of this strictly periodic shape function on the scattering matrix element, we can largely refer to the discussion of Sect. D.2. In view of the in-depth discussion given there, we will not recover the somewhat lengthy exact expression of the monochromatic scattering matrix element. Instead we are going to give merely a concise overview over the essential features of the monochromatic limit. As a monochromatic plane wave field is stretched over all space-time, the mentioned essence of the monochromatic limit will be found in the energy momentum conservation laws resulting from its analysis. One again finds the electron momenta, entering the energy momentum conservation law of the overall process, to be replaced by their dressed counterparts $q^\mu = p^\mu - \left(e^2 A^2/4\,(pk_L)\right) k_L^\mu$. Analogous to the discussion of Sect. D.2 we employ the generating function of the Bessel functions given in Eqs. (2.116) and (2.117) to obtain a Fourier expansion of the dynamic integrals. For the univariate integrals the resulting expansion is obtained analogous to Eq. (D.8) in the form

$$\begin{aligned} f_i^{\text{m.c.}} &= \sum_{n=-\infty}^{\infty} \int dx^- \; C_{i,n}\, \mathrm{e}^{-\mathrm{i}(q_i^+ + nk_L^+ - k_1^+ - k_2^+ - q_f^+)} \\ &= 2\pi \sum_{n=-\infty}^{\infty} C_{i,n}\, \delta\left(q_i^+ + nk_L^+ - k_1^+ - k_2^+ - q_f^+\right). \end{aligned} \tag{E.12}$$

The same phase contribution to the scattering matrix element thus features the four dimensional energy-momentum conservation, well known from the monochromatic case

$$S_{fi}^{\text{s.p.(m.c.)}} \propto \sum_{n=-\infty}^{\infty} \frac{C_{i,n}}{k_L(p_i - k_1)} \delta^{(4)}\left(q_i + nk_L - q_f - k_1 - k_2\right). \tag{E.13}$$

In this expression we note that the factor $((p_i - k_1)k_L)^{-1} = (p_t k_L)^{-1}$ stems from the dressed electron propagator and vanishes. We thus conclude that, also in the monochromatic limit, there are no divergences introduced into the same phase partial amplitude. As mentioned in the beginning of Chap. 4, therefore this partial amplitude could be studied in already published purely monochromatic analyses [2, 3].

The explicitly diverging resonances of the NDCS amplitude in an monochromatic external field must then be due to the phase ordered partial amplitude of the scattering matrix element. To recover the energy momentum conservation laws of this partial process we will have to analyze the bivariate dynamic integrals f_{ij}. It is, however, impossible to obtain δ-functions from integrations over only the half space $x_1^- < x_2^-$. To avoid this shortcoming we adopt a procedure, similar to that of Sect. 4.4, and insert the inverse of Eq. (E.4) into their definition (4.13), which yields the expression

$$\begin{aligned} f_{ij}^{\text{m.c.}} =& -\mathrm{i} \sum_{n,n'=-\infty}^{\infty} \lim_{\epsilon\to 0} \int dx^- dy^- \frac{dq^+}{2\pi} \frac{C_{i,n'} C_{j,n}}{q^+ - \mathrm{i}\epsilon} \\ &\times \mathrm{e}^{-\mathrm{i}(q_i^+ + nk_L^+ - k_1^+ - q_t^+ - q^+)x^-} \mathrm{e}^{-\mathrm{i}(q_t^+ + q^+ + n'k_L^+ - k_2^+ - q_f^+)y^-} \\ =& -(2\pi\mathrm{i}) \sum_{n,n'=-\infty}^{\infty} \frac{C_{i,n'} C_{j,n}}{q_i^+ + nk_L^+ - k_1^+ - q_t^+} \delta\left(q_i^+ + (n+n')k_L^+ - k_1^+ - k_2^+ - q_f^+\right), \end{aligned} \tag{E.14}$$

where $q_t^\mu = p_t^\mu + m^2\xi^2/4(p_t k_L)k_L^\mu$ is the dressed transitional electron momentum, fulfilling the dressed mass shell condition $q_t^2 = m^{*2}$. Combining this expression with the perpendicular energy momentum conservation of Eq. (4.2), this implies that in the monochromatic limit the phase ordered partial process diverges, if it holds

$$\begin{aligned} q_i + nk_L &= k_1 + q_t \\ q_t + n'k_L &= k_2 + q_f. \end{aligned}$$

Due to the energy momentum conservation of Eq. (E.14) the above conditions obviously are equivalent. The found resonances are hit at the photon frequencies

$$\omega_1 = \frac{n(q_i k_L)}{(q_i + nk_L)n_1} \tag{E.15a}$$

$$\omega_2 = \frac{n'(q_f + k_2)k_L)}{(q_f n_2)}, \tag{E.15b}$$

which are the so-called Oleinik resonances, typical of any multiple process occurring in a monochromatic external field [2, 4–6]. We have accordingly found that the divergences arising when a monochromatic external field is considered are in fact arising from the phase ordered contribution. We stress that contrary to the singular structure found above for such a strictly periodic scattering field, the bivariate dynamic integrals f_{ij}, leading to this behaviour, are finite when the shape function is confined in time, i.e. $\psi_{\mathcal{A}}(\eta \to \pm\infty) = 0$. We thus state that the Oleinik resonances are merely due to the unphysical assumption of an infinitely extended external field and are absent when considering a realistic scenario.

References

1. J.M. Jauch, F. Rohrlich, *Theory of Photons and Electrons* (Springer, Heidelberg, 1976)
2. E. Lötstedt, U.D. Jentschura, Nonperturbative Treatment of Double Compton Backscattering in Intense Laser Fields. Phys. Rev. Lett. **103**, 110404 (2009)
3. E. Lötstedt, U.D. Jentschura, Correlated two-photon emission by transitions of Dirac-Volkov states in intense laser fields: QED predictions. Phys. Rev. A **80**, 053419 (2009)
4. V.P. Oleinik, Resonance effects in field of an intense laser beam. Sov. Phys. JETP **25**, 697 (1967)
5. K. Krajewska, Electron-positron pair creation and Oleinik resonances. Laser Phys **21**, 1275 (2011)
6. D. Seipt, B. Kämpfer, Two-photon Compton process in pulsed intense laser fields. Phys. Rev. D **85**, 101701 (2012)

Curriculum Vitae

Personal Data

K. Felix Mackenroth
c/o Max-Planck-Institut für Kernphysik
Saupfercheckweg 1 • 69117 Heidelberg
Germany • 06221/516176
felix.mackenroth@mpi-hd.mpg.de
Date of birth: August 1st 1984
Place of birth: Lübeck, Germany

Professional Experience

Post-Doc Work

since 12/2012 Post-Doc at the Max Planck Institute for Nuclear Physics (MPIK), Heidelberg

Education

12/2012 **PhD degree Heidelberg University**
summa cum laude

07/2009 **Diploma at Heidelberg University**
grade: **1.0** (*awarded with distinction*)
minors: Astrophysics, Strong-Field QED

2005 Intermediate Examination at **University Konstanz**
2003 **Abitur** at Kaiser-Karl-Schule, Itzehoe
Advanced Courses: Physics, English

K. F. Mackenroth, *Quantum Radiation in Ultra-Intense Laser Pulses*,
Springer Theses, DOI: 10.1007/978-3-319-07740-6,

Awards

- **Otto Hahn Medal 2013** of the Max Planck Society
- **Online Stipend** e-fellows.net (2005-2009)
- **Book price** of the German Physical Society (**DPG**) (2003)

Voluntary Service

- 2010/11 **PhD representative** of the Max-Planck-Institute for Nuclear Physics
- 2004/05 Student member of the **Physics faculty council** of the University Konstanz
- 2004/05 tutor at the students bike garage of the **University Konstanz**

Languages

Languages	German: Native English: Fluent

Practical Experience

Work as Research Assistant

2007/08 **MiniResearch**
University Heidelberg, IUP, Group Dr. Wagenbach (**Glaciology**)
Measuring the complex dielectric constant of ice

2006 **Undergraduate research assistant**
University Heidelberg, IWR, Group Prof. Hamprecht
Digital image processing *using MatLab*

Internships

2007 Six months **Industrial Internship** at **BASF AG**, Ludwigshafen (Germany)
Theoretical modelling and experimental investigation of heat conductance in high temperature reactors

07/2002 **Economy Summer School** *of the federal state of Schleswig-Holstein, Germany (WIWAG)*

Publications

Peer-Reviewed Journal Articles

A photon-photon collider in a vacuum hohlraum
O.J. Pike, F. Mackenroth, E.G. Hill and S.J. Rose
Nature Photonics **8**, 434 (2014). doi:10.1038/nphoton.2014.95

Novel aspects of radiation reaction in the ultrarelativistic quantum regime
F. Mackenroth, N. Neitz and A. Di Piazza
Plasma Phys. Control. Fusion **55**, 124018 (2013)
[chosen for LabTalk communication by Plasma Phys. Control. Fusion]

Nonlinear double Compton scattering in the ultrarelativistic quantum regime
F. Mackenroth and A. Di Piazza
Physical Review Letters **110**, 070402 (2013)

Time-dependent Kohn-Sham approach to quantum electrodynamics
M. Ruggenthaler, F. Mackenroth and D. Bauer
Physical Review A **84**, 042107 (2011)

Nonlinear Compton scattering in ultrashort laser pulses
F. Mackenroth and A. Di Piazza Physical Review A **83**, 032106 (2011)

Determining the Carrier-Envelope Phase of Intense Few-Cycle Laser Pulses
F. Mackenroth, A. Di Piazza and C.H. Keitel
Physical Review Letters **105**, 063903 (2010)

Unrefereed Conference Proceedings

Strong-field QED in intense laser fields
F. Mackenroth, M. Ruf, B. King, A. Di Piazza, C. Müller and C.H. Keitel
SPIE Proceedings **7994**, 799422 (2010)